UNDERGROUND CLINICAL VIGNETTES

··

ANATOMY

Classic Clinical Cases for
USMLE Step 1 Review [100 cases, 2nd ed]

VIKAS BHUSHAN, MD
University of California, San Francisco, Class of 1991
Series Editor, Diagnostic Radiologist

CHIRAG AMIN, MD
University of Miami, Class of 1996
Orlando Regional Medical Center, Resident in Orthopaedic Surgery

TAO LE, MD
University of California, San Francisco, Class of 1996
Yale-New Haven Hospital, Resident in Internal Medicine

PARAG MATHUR
Mayo School of Medicine, Class of 2001

JOSE M. FIERRO, MD
La Salle University, Mexico City
Brookdale University Hospital, New York, Intern in Medicine/Pediatrics

HOANG NGUYEN
Northwestern University, Class of 2000

Editorial Office:
Commerce Place, 350 Main Street, Malden, Massachusetts 02148, USA

Distributed by Blackwell Science, Inc.:

USA

Blackwell Science, Inc.
Commerce Place
350 Main Street
Malden, Massachusetts 02148
(Telephone orders: 800-215-1000 or
781-388-8250;
fax orders: 781-388-8270)

Canada

Login Brothers Book Company
324 Saulteaux Crescent
Winnipeg, Manitoba, R3J 3T2
(Telephone orders: 204-224-4068;
Telephone: 800-665-1148; fax: 800-
665-0103)

Australia

Blackwell Science Pty, Ltd.
54 University Street
Carlton, Victoria 3053
(Telephone orders: 03-9347-0300;
fax orders: 03-9349-3016)

Outside North America and Australia

Blackwell Science, Ltd.
c/o Marston Book Services, Ltd.
P.O. Box 269
Abingdon
Oxon OX14 4YN
England
(Telephone orders: 44-01235-465500;
fax orders: 44-01235-465555)

ISBN: 1-890061-19-0

Editor: Andrea Fellows
Typesetter: Vikas Bhushan using MS Word97
Printed and bound by Capital City Press

Printed in the United States of America
99 00 01 02 5 4 3

Contributors

SAMIR MEHTA
Temple University, Class of 2000

ALEA EUSEBIO
UCLA School of Medicine, Class of 2000

AARCHAN JOSHI, MD
UCLA Jules Stein Eye Institute

VIPAL SONI
UCLA School of Medicine, Class of 1999

DIEGO RUIZ
UCSF School of Medicine, Class of 1999

Acknowledgments

· ·

Throughout the production of this book, we have had the support of many friends and colleagues. Special thanks to our business manager, Gianni Le Nguyen. For expert computer support, Tarun Mathur and Alex Grimm. For design suggestions, Sonia Santos and Elizabeth Sanders.

For editing, proofreading, and assistance across the vignette series, we collectively thank Carolyn Alexander, Henry E. Aryan, Natalie Barteneva, Sanjay Bindra, Julianne Brown, Hebert Chen, Arnold Chin, Yoon Cho, Karekin R. Cunningham, A. Sean Dalley, Sunit Das, Ryan Armando Dave, Robert DeMello, David Donson, Alea Eusebio, Priscilla A. Frase, Anil Gehi, Parul Goyal, Alex Grimm, Tim Jackson, Sundar Jayaraman, Aarchan Joshi, Rajni K. Jutla, Faiyaz Kapadi, Aaron S. Kesselheim, Sana Khan, Andrew Pin-wei Ko, Warren S. Krackov, Benjamin H.S. Lau, Scott Lee, Warren Levinson, Eric Ley, Ken Lin, Samir Mehta, Gil Melmed, Joe Messina, Vivek Nandkarni, Deanna Nobleza, Darin T. Okuda, Adam L. Palance, Sonny Patel, Ricardo Pietrobon, Riva L. Rahl, Aashita Randeria, Marilou Reyes, Diego Ruiz, Anthony Russell, Sanjay Sahgal, Sonal Shah, John Stulak, Lillian Su, Julie Sundaram, Rita Suri, Richa Varma, Amy Williams, Ashraf Zaman and David Zipf. Please let us know if your name has been missed or mispelled and we will be happy to make the change in the next edition.

Table of Contents

. . . .

CASE	SUBSPECIALTY	NAME
40	Neurology	Conductive Deafness
41	Neurology	Epiphyseal Separation with Ulnar Nerve
42	Neurology	Facial Nerve Injury
43	Neurology	Femoral Nerve Palsy
44	Neurology	Herniation of Intervertebral Disk
45	Neurology	Hypoglossal Nerve Palsy
46	Neurology	Klumpke's Palsy
47	Neurology	Medial Medullary Syndrome
48	Neurology	Obstructive Hydrocephalus
49	Neurology	Obstructive Sleep Apnea
50	Neurology	Parkinson's Disease
51	Neurology	Spina Bifida with Myelomeningocele
52	Neurology	Subarachnoid Hemorrhage with CN III
53	Neurology	Tabes Dorsalis
54	Neurology	Trigeminal Neuralgia
55	Neurology	Wallenberg's Syndrome
56	Neurology	Wernicke's Aphasia
57	OB/GYN	Caput Succedaneum
58	OB/GYN	Gartner's Duct Cyst
59	OB/GYN	Pudendal Nerve Block
60	OB/GYN	Ruptured Ectopic Pregnancy
61	OB/GYN	Uterine Prolapse with Cystocele
62	Orthopedics	Acromioclavicular Dislocation
63	Orthopedics	Ankle Sprain
64	Orthopedics	Anterior Compartment Syndrome
65	Orthopedics	Boxer's Fracture
66	Orthopedics	Carpal Tunnel Syndrome
67	Orthopedics	Cervical Rib
68	Orthopedics	Combined Knee Injury
69	Orthopedics	Erb's Palsy
70	Orthopedics	Femoral Neck Fracture
71	Orthopedics	Fracture of Clavicle
72	Orthopedics	Lateral Epicondylitis
73	Orthopedics	Legg–Calve–Perthes Disease
74	Orthopedics	Long Thoracic Nerve Injury
75	Orthopedics	Median Nerve Palsy (Noncarpal)
76	Orthopedics	Osgood–Schlatter's Disease
77	Orthopedics	Pelvic Fracture
78	Orthopedics	Radial Head Subluxation
79	Orthopedics	Radial Nerve Palsy
80	Orthopedics	Scaphoid Fracture
81	Orthopedics	Shoulder Dislocation
82	Orthopedics	Slashing of Wrists
83	Orthopedics	Temporomandibular Joint Dislocation
84	Orthopedics	Trendelenburg Gait
85	Surgery	Direct Inguinal Hernia
86	Surgery	Duodenal Atresia
87	Surgery	Indirect Inguinal Hernia

Preface to the Second Edition

. . . .

We are very pleased with the overwhelmingly positive reception of the first edition of our *Underground Clinical Vignettes* series. In the second editions we have fine-tuned nearly every case by incorporating corrections, enhancements and clarifications. These were based on feedback from the several thousand students who used the first editions.

We implemented two structural changes upon the request of many students:

♦ bi-directional cross-linking to appropriate High Yield Facts in the 1999 edition of *First Aid for the USMLE Step1* (Appleton & Lange);

♦ case names have been moved to the bottom of the page and obvious references to the case name within the case description have been removed.

With this, we hope they'll emerge as a unique and well-integrated study tool that provides compact clinical correlations to basic science information.

We invite your corrections and suggestions for the next edition of this book. For the first submission of each factual correction or new vignette, you will receive a personal acknowledgement and a free copy of the revised book. We prefer that you submit corrections or suggestions via electronic mail to vbhushan@aol.com. Please include "Underground Vignettes" as the subject of your message. If you do not have access to e-mail, use the following mailing address: S2S Medical Publishing, 1015 Gayley Ave, Box 1113, Los Angeles, CA 90024 USA.

Preface to the First Edition

. . . .

This series was developed to address the increasing number of clinical vignette questions on the USMLE Step 1 and Step 2. It is designed to supplement and complement *First Aid for the USMLE Step 1* (Appleton & Lange).

Each book uses a series of approximately 100 "**supra-prototypical**" cases **as a way to condense testable facts and associations**. The clinical vignettes in this series are designed to incorporate as many testable facts as possible into a cohesive and memorable clinical picture. The vignettes represent composites drawn from general and specialty textbooks, reference books, thousands of USMLE style questions and the personal experience of the authors and reviewers.

Although each case tends to present all the signs, symptoms, and diagnostic findings for a particular illness, **patients generally will not present with such a "complete" picture either clinically or on the Step 1 exam**. Cases are not meant to simulate a potential real patient or an exam vignette. All the **boldfaced "buzzwords" are for learning purposes** and are not necessarily expected to be found in any one patient with the disease.

Definitions of selected important terms are placed within the vignettes in (= SMALL CAPS) in parentheses. Other parenthetical remarks often refer to the pathophysiology or mechanism of disease. The format should also help students learn to present cases succinctly during oral "bullet" presentations on clinical rotations. The cases are meant to be read as a condensed review, not as a primary reference.

The information provided in this book has been prepared with a great deal of thought and careful research. This book should not, however, be considered as your sole source of information. Corrections, suggestions and submissions of new cases are encouraged and will be acknowledged and incorporated in future editions.

Abbreviations

. . . .

ABGs – arterial blood gases
ACL – anterior cruciate ligament
AIDS – acquired immunodeficiency syndrome
ALT – alanine transaminase
Angio – angiography
AP –anteroposterior
AST – aspartate transaminase
AV – arteriovenous
BE – barium enema
BP – blood pressure
BUN – blood urea nitrogen
CBC – complete blood count
CK – creatine kinase
CN – cranial nerve
CP – cerebellopontine angle
CSF – cerebrospinal fluid
CT – computerized tomography
CXR – chest x-ray
DES - diethylstilbestrol
DVT – deep venous thrombosis
ECG – electrocardiography
Echo - echocardiography
EEG – electroencephalography
EGD – esophagogastroduodenoscopy
EMG – electromyography
ERCP – endoscopic retrograde cholangiopancreatography
FNA – fine needle aspiration
FTA-ABS – fluorescent treponemal antibody absorption
HEENT – head, eyes, ears, nose, and throat
HIDA – hepatoiminodiacetic acid [scan]
HPI – history of present illness
HR – heart rate
ID/CC – identification and chief complaint
IMA – inferior mesenteric artery
IVP – intravenous pyelography
JVP – jugular venous pressure
KUB – kidneys/ureter/bladder
LP – lumbar puncture
Lytes – electrolytes
Mammo – mammography
MAO – monoamine oxidase
MR – magnetic resonance [imaging]
NF – neurofibromatosis
NSAID – nonsteroidal anti-inflammatory drug
Nuc – nuclear medicine
PA – posteroanterior
PBS – peripheral blood smear
PE – physical exam
PET – positron emission tomography

Abbreviations - continued

PFTs – pulmonary function tests
PPD – purified protein derivative
PT – prothrombin time
PTT – partial thromboplastin time
PVC – premature ventricular contraction
SBFT – small bowel follow-through [barium study]
SMA – superior mesenteric artery
TMJ – temporomandibular joint
UA – urinalysis
UGI – upper GI [barium study]
US – ultrasound
V/Q – ventilation perfusion
VDRL – Venereal Disease Research Laboratory
VS – vital signs
VSD – ventricular septal defect
WBC – white blood cell
XR – x-ray

ID/CC	A 29-year-old male comes to the medical clinic because of **palpitations**, weakness, and **fatigue** that does not allow him to walk more than five blocks, together with **coldness of his right foot.**
HPI	He underwent surgery four weeks ago for a penetrating stab-wound **injury in his right groin** that he sustained during a fight.
PE	VS: **marked tachycardia**. PE: **continuous murmur** and easily palpable **thrill** over area of wound; skin over wound warm to touch; right foot cold to touch with **diminished pulse;** tachycardia diminished when pressure applied to site of fistula (= BRANHAM'S SIGN).
Labs	CBC/Lytes: normal. LFTs, glucose, BUN, creatinine normal.
Imaging	MR/Angio: large AV connection (fistula) in groin area with significant diversion of blood flow. US: color flow doppler shows rainbow-colored **turbulence** in fistula; high-velocity and arterialized (pulsatile) waveform in draining vein.
Gross Pathology	Abnormal communication between artery and vein, in this case as a result of a penetrating injury.
Micro Pathology	N/A
Treatment	Surgical repair if symptomatic and large; angiographic embolization if smaller. Ultrasound-guided direct compression is sometimes an option.
Discussion	May clinically present as high-output cardiac failure. Iatrogenic AV fistulas may be seen after arteriography.

ARTERIOVENOUS FISTULA

ID/CC	A 42-year-old female presents with progressive <u>shortness of breath on exertion</u> and <u>palpitations.</u>
HPI	The patient has been symptom free until now.
PE	VS: irregularly irregular pulse. PE: left parasternal heave; grade III/VI <u>systolic ejection flow murmur</u> in left second intercostal space; <u>widely split, fixed S2</u> (does not change with breathing).
Labs	ECG: atrial fibrillation; RSR pattern in right precordial leads; right-axis deviation (right ventricular hypertrophy).
Imaging	CXR: dilated proximal pulmonary arteries; <u>increased pulmonary vascularity;</u> enlarged right atrium and right ventricle; small aortic knob. Echo: <u>paradoxical septal movement;</u> left-to-right flow. Cardiac catheterization confirmatory.
Gross Pathology	The most common form is in the **midseptum,** in the area of the foramen ovale (= OSTIUM SECUNDUM); lower septum (= OSTIUM PRIMUM) associated with AV valve anomalies (most common in Down's); upper septum (= SINUS VENOSUS) associated with anomalous pulmonary venous return.
Micro Pathology	N/A
Treatment	Surgical or interventional angiographic closure of defect with prosthetic patch. Operative repair is recommended in all symptomatic patients with ostium secundum defects regardless of size of defect.
Discussion	Oxygenated blood from the left atrium passes into the right atrium, increasing right ventricular output and pulmonary flow. **Acyanotic** (left-to-right shunt); the most common congenital heart disease in adults. Sequelae of untreated atrial septal defects include **paradoxic emboli, infective endocarditis,** and **congestive heart failure.**

ATRIAL SEPTAL DEFECT

ID/CC	A 25-year-old male postal worker who was ~~stabbed in the chest~~ during a mugging is brought to the emergency room in a ~~semiconscious state, gasping for air~~ (= DYSPNEA).
HPI	The knife penetrated the thoracic wall at the level of the fourth intercostal space along the left sternal border.
PE	VS: ~~hypotension (BP 90/40) that does not respond to rehydration;~~ inspiratory lowering of systolic BP by > 10 mmHg (= PULSUS PARADOXUS). PE: <u>increase in venous pressure</u> with inspiratory filling of neck veins during ~~inspiration (= KUSSMAUL'S SIGN);~~ during drawing of venous blood, syringe filled spontaneously (due to increased venous pressure); **apical impulse diminished; heart sounds seem distant; patient also cyanotic.**
Labs	ECG: reduced voltage.
Imaging	CXR: cardiomegaly, but with acute hemopericardium, the heart shadow may not enlarge; thus diagnosis is clinical. Echo: ~~pericardial fluid;~~ diastolic collapse of right ventricle and atria.
Gross Pathology	**Blood** from sites of injury **fills pericardial sac,** causing compression of all heart chambers and preventing venous return, heart filling, and arterial outflow.
Micro Pathology	N/A
Treatment	**Immediate pericardiocentesis** and subsequent operative thoracotomy and pericardial decompression with repair of laceration.
Discussion	Unlike this case, the majority of patients with penetrating chest trauma will have a pneumothorax or hemothorax. The triad of Beck (**hypotension, distant heart sounds, and increased venous pressure**) is characteristic of cardiac tamponade.

CARDIAC TAMPONADE

ID/CC	An 18-year-old white <u>**male**</u> is found during a military physical to have <u>**high blood pressure.**</u>
HPI	The patient denies a history of any major illness.
PE	VS: <u>**BP in arms significantly greater than BP in legs**</u> (BP right arm 180/110, left arm 190/110; BP in legs 110/70). PE: <u>**femoral pulses diminished** and **delayed;**</u> harsh <u>**systolic ejection murmur**</u> heard between shoulder blades.
Labs	ECG: left ventricular hypertrophy.
Imaging	CXR: may demonstrate <u>**rib notching**</u> (due to collateral circulation through intercostal arteries); poststenotic dilatation of aorta. Echo/MR/Angio: diagnostic.
Gross Pathology	In 95% of cases, the coarctation <u>(**narrowing**)</u> is distal to left subclavian artery. <u>The infantile type is proximal to the ligamentum arteriosum; the adult type is distal.</u>
Micro Pathology	N/A
Treatment	Surgical repair or balloon angioplasty.
Discussion	**Acyanotic;** twice as common in males as in females; frequently associated with ventricular septal defect, patent ductus arteriosus, and bicuspid aortic valve. The most common surgically correctable causes of **secondary hypertension** include Conn's syndrome (aldosterone-producing adrenocortical adenoma), renal artery stenosis, coarctation of the aorta, and pheochromocytoma. Turner's syndrome (45, XO) is associated with an increased incidence of coarctation of the aorta. **FIRST AID** p.216

. . . .

COARCTATION OF THE AORTA

ID/CC   A 57-year-old male is seen by the resident on call because he complains of **pain in the groin** area and **coldness in the right foot.**

HPI   He underwent **coronary angiography** that morning for evaluation of coronary artery disease.

PE   VS: tachycardia (HR 98); hypotension (BP 90/60); no fever. PE: **pallor;** right groin examination discloses marked **swelling and deformation** at site of femoral artery puncture with **skin discoloration (ecchymosis)** to middle third of thigh anteriorly and posteriorly (patients are anticoagulated for angiography); **peripheral pulses** present but **diminished.**

Labs   CBC: low hematocrit. Prolonged clotting time, PT, and PTT.

Imaging   US: shows hematoma (no flow, nonpulsatile) and excludes pseudoaneurysm.

Gross Pathology   Large subcutaneous hematoma at site of puncture, causing compression of femoral artery.

Micro Pathology   N/A

Treatment   Evacuation of hematoma, compressive bandage, and drainage.

Discussion   The external iliac artery passes in close proximity to the inguinal ligament, where it is susceptible to injury during hernia repair. Distal to this landmark, the artery changes its name to the common femoral artery, which is the main blood supply to the leg. In the groin, the neurovascular bundle supplying the lower extremity consists of the **femoral vein, artery, and nerve,** in that order, from **medial to lateral.**

FEMORAL HEMATOMA

ID/CC	An 8-year-old female with a history of ~~recurrent pneumonia and low exercise tolerance~~ is referred to a pediatric cardiologist for evaluation.
HPI	The child was born prematurely and has a history of recurrent respiratory tract infections; her ~~mother had rubella during her pregnancy.~~
PE	Delayed growth and development (fifth percentile); ~~wide pulse pressure;~~ prominent carotid pulsation; increased JVP; ~~continuous "machinery" murmur~~ with systolic accentuation and thrill at second intercostal space at left parasternal border; increased intensity of apical impulse.
Labs	ECG: increased voltage of R in V_5-V_6 and S in V_1-V_2; left axis deviation (left ventricular hypertrophy).
Imaging	CXR: enlarged cardiac shadow with increased pulmonary blood flow (left atrium, left ventricle, aorta, and pulmonary artery). Echo: doppler flow mapping confirmatory. Angio: definitive.
Gross Pathology	Patent ductus arteriosus (PDA) **connects the aorta and left pulmonary artery just distal to the origin of the left subclavian artery.**
Micro Pathology	N/A
Treatment	Surgical or catheter (umbrella) closure. **Indomethacin** alone may be successful in closing the PDA in neonates (due to **inhibition of synthesis of prostaglandin E2, which normally keeps the ductus arteriosus open prior to birth).**
Discussion	**Acyanotic;** seen in association with congenital rubella. **FIRST AID** p.217

PATENT DUCTUS ARTERIOSUS

ID/CC	A 6-year-old child is referred to a pediatric cardiologist for evaluation of **dyspnea on exertion.**
HPI	Since birth, he has had several-minute-long **"blue spells"** during which he becomes hyperpneic, cyanotic, and restless; at times he has also lost consciousness. He has been observed to assume a **squatting position in order to relieve dyspnea** due to physical effort (increases venous return to right heart and pulmonary flow).
PE	Delayed growth; central **cyanosis**; grade III **clubbing** of fingers; systolic thrill palpable along left parasternal border (due to ventricular septal defect); systolic murmur best heard at third left intercostal space (pulmonary stenosis); murmur disappears during cyanotic spell (no blood flow through valve); single second heart sound (only A2; soft, inaudible P2 due to pulmonary stenosis).
Labs	CBC: polycythemia. ABGs: hypoxemia (PaO_2 72%). ECG: right axis deviation; evidence of **right ventricular hypertrophy** and right atrial dilatation.
Imaging	CXR: concavity in region of main pulmonary artery; right ventricular enlargement (= BOOT-SHAPED HEART); **diminished pulmonary vascularity.** Echo: shows all four gross findings. Cardiac catheterization confirmatory.
Gross Pathology	Four defects noted: (1) large **ventricular septal defect;** (2) **right ventricular outflow obstruction** (pulmonary valve stenosis); (3) **right ventricular hypertrophy;** (4) **"overriding" large ascending aorta.**
Micro Pathology	N/A
Treatment	Palliative shunt or corrective open heart surgery.
Discussion	A life-threatening condition; a common cause of cyanosis in childhood. Symptomatology is directly proportional to the amount of right ventricular outflow obstruction. Caused by an **embryologic defect** that causes **anterosuperior displacement of the infundibular septum,** resulting in unequal division of the aorta and pulmonary artery. **FIRST AID** p.216

TETRALOGY OF FALLOT

ID/CC	A 2-month-old female presents with **dyspnea,** feeding difficulties, **poor growth,** and profuse perspiration.
HPI	The child had pneumonia when she was seven days old, at which time her parents were informed of a congenital heart murmur.
PE	VS: pulse normal. PE: no edema, cyanosis, or clubbing; **palpable parasternal heave;** apical thrust and systolic thrill; loud P2 component of S2; harsh **pansystolic murmur heard best over left lower sternal border;** short apical diastolic rumble (due to increased flow across mitral valve).
Labs	ECG: **biventricular hypertrophy** with peaked P waves due to right atrial hypertrophy (pulmonary artery hypertension).
Imaging	CXR: cardiomegaly (all chambers and pulmonary artery except right atrium) with increased pulmonary vascularity. Echo: large ventricular septal defect (VSD); doppler shows **left-to-right** direction of shunt (left ventricle to right ventricle).
Gross Pathology	**Failure of fusion of interventricular septum with aortic septum.** Four types of VSDs: perimembranous (80%), muscular (10%), supracristal (5%), and endocardial cushion defect (5%).
Micro Pathology	N/A
Treatment	Surgical closure of large VSDs should be performed before appearance of irreversible pulmonary vascular hypertension.
Discussion	**Acyanotic.** A common cardiac malformation that accounts for 25% of all cases of congenital heart disease. Small defects may close spontaneously. The development of pulmonary vascular hypertension may lead to reversal of the shunt (into right-to-left) and cyanosis (= EISENMENGER'S SYNDROME).

VENTRICULAR SEPTAL DEFECT

ID/CC	A newborn male is evaluated by a neonatologist because of <u>cyanosis.</u>
HPI	The child presents with cyanosis that <u>**increases with feeding**</u> (while the child uses the mouth for eating, no air goes in the lungs) <u>**and is relieved with crying.**</u>
PE	Well developed and hydrated; heart sounds normal; no murmurs; lungs clear; no increase in JVP; resident was <u>unable to pass a catheter through the nose</u> (diagnostic feature).
Labs	Lab studies and neonatal screen normal.
Imaging	CT: confirmatory.
Gross Pathology	N/A
Micro Pathology	N/A
Treatment	Surgical correction.
Discussion	Newborns are **obligate nose breathers,** so patients with choanal atresia cannot inhale enough air and thus become cyanotic. When the child cries, air is breathed into his lungs via the mouth, correcting the cyanosis. A normal choana allows communication between the nasal fossa and the nasopharynx.

CHOANAL ATRESIA

ID/CC	An 18-year-old college freshman suddenly collapses in the middle of a dinner at his fraternity house; shortly thereafter his face turns blue (= CYANOSIS) and he struggles desperately to breathe.
HPI	He had been drinking heavily most of the afternoon while celebrating his school's football victory (and thus was less able to chew his food properly, had decreased sensation in his mouth, exercised less caution, and had impairment of the cough reflex). He was also laughing heartily while eating.
PE	VS: tachycardia. PE: in acute distress, clutching throat with both hands; cyanotic; sweaty, with inspiratory stridor and high-pitched expiratory sounds while attempting to breathe; altered level of consciousness; piece of meat lodged at inlet of larynx; spasm of laryngeal muscles. Patient stopped breathing and collapsed.
Labs	N/A
Imaging	XR: lateral view of neck may show foreign body causing airway obstruction (obtained in stable patients with partial obstruction).
Gross Pathology	N/A
Micro Pathology	N/A
Treatment	Manual removal of obstructing foreign body, rapid back blows, Heimlich maneuver, or emergency cricothyroidotomy (incision through cricoid ligament inferior to thyroid cartilage and superior to cricoid cartilage).
Discussion	Prevention is the key, mainly in children; teaching the Heimlich maneuver to laypeople has saved the lives of many "café coronary" victims.

. . . .

CHOKING

ID/CC   A 46-year-old female on her third postoperative day rings the nurse because of the development of numbness around her mouth as well as a tingling sensation in her legs and fingertips.

HPI   The patient has spastic contracture of the feet and wrists in outward rotation and flexion, with her fingertips touching each other (= CARPOPEDAL SPASM). She had just had a total thyroidectomy.

PE   VS: normal. PE: surgical wound healed; no signs of infection or hematoma formation; contraction of facial muscles when tapping facial nerve anterior to ear (= POSITIVE CHVOSTEK'S SIGN); carpal spasm after occlusion of brachial artery with BP cuff (= TROUSSEAU'S SIGN); abduction and flexion of foot when peroneal nerve is tapped (= POSITIVE PERONEAL SIGN) (all signs of hypocalcemia).

Labs   CBC: normal. Hypocalcemia. ABGs: normal. ECG: normal.

Imaging CXR/KUB: normal.

Gross Pathology Resected thyroid tissue shows anaplastic carcinoma with incipient invasion into trachea (a small square of anterior wall was resected); on careful examination, all four parathyroid glands found to be deeply adherent to the thyroid; no evidence of nerve tissue (laryngeal nerve).

Micro Pathology N/A

Treatment Calcium supplements.

Discussion The parathyroid glands are the embryologic derivatives of the dorsal endoderm of the third and fourth branchial pouches. The glands may be found anywhere from the superior mediastinum to the carotid bifurcation but are usually located on the posterior aspect and in close proximity to or embedded in the thyroid gland. Usually there are two superior and two inferior glands, but supernumerary and absent glands are not uncommon.

IATROGENIC HYPOPARATHYROIDISM

ID/CC	During her first postop visit, a 23-year-old female complains to her surgeon of **hoarseness.**
HPI	She had just undergone a total **thyroidectomy** for papillary thyroid cancer.
PE	VS: normal. PE: surgical wound healed; no signs of infection or hematoma formation; no focal neurologic deficits; Chvostek's and Trousseau's signs absent (checking for hypocalcemia due to possible parathyroid removal); **unilateral vocal cord palsy with hoarseness** noted.
Labs	CBC/Lytes: normal. Glucose, BUN, creatinine normal; no hypocalcemia.
Imaging	N/A
Gross Pathology	The nerve was damaged during thyroid surgery while suturing the blood vessels of the inferior pole of the thyroid.
Micro Pathology	N/A
Treatment	Speech therapy.
Discussion	There are two recurrent laryngeal nerves (also called inferior laryngeal), both of which are branches of the **vagus nerve;** they are called recurrent because they **loop around the subclavian artery on the right and the aortic arch on the left** before ascending in the tracheoesophageal groove in close proximity to the thyroid gland to end up in the larynx. If the left recurrent laryngeal nerve is involved, one should consider mass lesions such as enlarged lymph nodes in the **aorticopulmonary window.**

ID/CC A 54-year-old male complains of acute pain on the left side of his face whenever he eats accompanied by swelling of the same side of his face (due to trapping of saliva in the parotid duct); he also complains of having expelled "sandy" material with his saliva.

HPI The patient has had frequent bouts of **infectious parotitis** (= CHRONIC SIALADENITIS).

PE Firm, round **mass palpable** below zygomatic process of temporal bone (stone in parotid duct).

Labs N/A

Imaging CT or Sialography: irregularly enlarged Stensen's duct; may visualize stone.

Gross Pathology Stones composed of a mucinous core surrounded by calcium and phosphate salt deposition.

Micro Pathology N/A

Treatment Have patient **suck on lemon** to attempt stone expulsion through increased salivation. Removal of stones (= LITHOTOMY) by duct dilatation or surgical gland removal.

Discussion All the salivary glands and ducts may present with stone formation (= SIALOLITHIASIS); the condition is frequently associated with chronic infection of the glands.
Approximately 80% of salivary gland stones are found in the **submandibular gland** (Wharton's duct).

ID/CC	A 19-year-old woman presents with a **painless swelling just beneath her hyoid bone.**
HPI	The swelling has been getting larger over the past several weeks but has not been painful.
PE	Rounded, **midline**, well-demarcated, painless, **fluid-filled mass** that is not fixed and **moves superiorly when patient swallows** (vs. dermoid cysts, which do not move); no other neck masses or lymphadenopathy present.
Labs	Basic lab work and thyroid function tests normal.
Imaging	XR-Lateral Neck: may see mass composed of soft tissue with no calcification. Nuc: radioactive iodine may localize in cyst if cyst contains functioning thyroid tissue.
Gross Pathology	N/A
Micro Pathology	N/A
Treatment	Surgical removal of thyroglossal duct, cyst, and midportion of hyoid after confirming presence of adequate-functioning thyroid tissue elsewhere.
Discussion	Cysts may arise from the remnant of the thyroglossal duct, an embryologic structure formed during **migration of the thyroid from the base of the tongue (at the foramen cecum) to its final position in the neck.** They frequently become infected. The foramen cecum is the normal remnant of the thyroglossal duct. **FIRST AID** p.94

ID/CC	A 10-year-old boy is brought to the pediatrician complaining of **high fever, sore throat,** earache, **swollen glands,** and productive, greenish-white, blood-tinged sputum.
HPI	His mother states that the boy has had **recurrent bouts of sore throat** several times a year for the past five years, each time treated effectively with antibiotics.
PE	Mouth partially open (swollen pharyngeal tonsil obstructs nasopharyngeal isthmus); **tonsils markedly enlarged, hyperemic, and cryptic with spotted areas of pus;** inflammation of torus tubarius (protects opening of eustachian tube; auditory meatus is immediately anterior and inferior to pharyngeal tonsil, and infection of pharyngeal tonsils spreads up auditory tube, causing otitis media).
Labs	CBC: neutrophilic leukocytosis; **Antistreptolysin titer (ASO) high; throat culture** shows **beta-hemolytic streptococcus.**
Imaging	XR-Lateral Neck: thickened retropharyngeal prevertebral tissue. CT: pharyngeal or retropharyngeal mass (abscess) may be present.
Gross Pathology	N/A
Micro Pathology	N/A
Treatment	Penicillin. Retropharyngeal abscess is a serious complication that requires drainage. Evaluate for tonsillectomy.
Discussion	Tonsillectomy is performed less frequently now than a decade ago; nevertheless, an evaluation must be done weighing surgical risks with those of recurrent beta-streptococcal infections and possible rheumatic fever. Waldeyer's ring consists of the nasopharyngeal tonsils, the palatine tonsils, and the lingual tonsils.

. . . .

TONSILLITIS

ID/CC A 35-year-old woman is brought to the emergency room by ambulance because of the sudden appearance of **severe retrosternal pain** with radiation to the back and abdomen along with dyspnea; the pain appeared **after vigorous vomiting.**

HPI She suffers from episodes of binge eating and self-induced vomiting (= BULIMIA).

PE VS: **tachycardia** (HR 110); mild hypotension (BP 100/65); no fever. PE: in acute distress; complains of severe chest pain; no heart murmurs; left lung field **hypoaerated** (due to pneumothorax); crackling sound heard over precordium (= HAMMAN'S SIGN OF PNEUMOMEDIASTINUM).

Labs CBC: leukocytosis. Amylase elevated.

Imaging UGI: **extravasation of contrast** into mediastinum. CXR/CT: left pleural effusion and hydropneumothorax; **mediastinal emphysema.** Esophagoscopy: **complete rupture of esophageal wall.**

Gross Pathology All layers of the esophagus are torn completely in **posterior lateral wall of esophagus on left side** (vs. Mallory–Weiss tear of only superficial esophageal layers; presents as postemetic bleeding).

Micro Pathology N/A

Treatment Broad-spectrum antibiotics, chest tube and surgical repair.

Discussion Postemetic rupture of the esophageal wall (= BOERHAAVE'S SYNDROME) is usually seen following protracted and forceful vomiting of solid food; it is common in **alcoholics, bulimics, and pregnant women** and in any condition that increases intra-abdominal pressure. The esophagus has **three anatomic constrictions:** the cardiac (the most common site of rupture), the aortic arch, and the cricopharyngeal.

BOERHAAVE'S SYNDROME

ID/CC	A full-term, **3-week-old** male is brought to his family physician for his second well-baby visit, at which time the physician notices that the infant is **jaundiced** (jaundice did not start immediately after birth, as is the case with physiologic jaundice).
HPI	On directed questioning, the mother also reports that he has **dark urine** staining his diaper along with passage of "**clay-colored**" (= ACHOLIC) **stools** (due to obstructive jaundice).
PE	**Icteric skin and sclera;** firm mild hepatosplenomegaly; no signs of portal hypertension or liver failure.
Labs	**Direct hyperbilirubinemia; increased alkaline phosphatase,** ALT, and AST; **low serum albumin;** increased globulin; **lack of urobilinogen in urine.**
Imaging	US: normal. Nuc-HIDA Scan: unimpaired liver uptake with **absent excretion into intestine over 24 hours.**
Gross Pathology	Liver increased in size with green-colored, granular surface; periportal fibrosis if long-standing; **extrahepatic bile duct consists of fibrous cords with no lumen** (atretic bile ducts).
Micro Pathology	Liver biopsy shows bile duct proliferation with dilatation of canaliculi and presence of inspissated bile plugs.
Treatment	Surgery before two months of age to prevent liver damage; Kasai procedure to directly attach bowel to surface of liver; liver transplant.
Discussion	The most common cause of **persistent jaundice in infancy;** associated with the presence of more than one spleen (= POLYSPLENIA). Differential diagnosis includes choledochal cyst (mass usually palpable), alpha-1-antitrypsin deficiency, and neonatal hepatitis. If **long-standing, liver cirrhosis** will develop; other complications include chronic cholangitis, fat-soluble vitamin deficiencies, and portal hypertension.

. . . .

CONGENITAL BILIARY ATRESIA

ID/CC A neonatologist is called into the nursery for an emergency; a **newborn** baby girl has become **dyspneic** and **turned blue** (= CYANOTIC) upon her arrival from the delivery room.

HPI The patient is the product of normal delivery. Her mother **did not receive any prenatal care.**

PE Full-term female baby; cyanosis; **severe dyspnea** with obvious **intercostal retractions;** small and scaphoid abdomen; **absent breath sounds** and **positive peristaltic bowel sounds in left chest;** heart sounds heard best over right hemithorax (due to cardiomediastinal shift).

Labs N/A

Imaging CXR: **coils of air-filled stomach or bowel seen in left hemithorax,** displacing heart to right side. Prenatal diagnosis can be made by ultrasonography.

Gross Pathology Left pulmonary hypoplasia; left posterolateral congenital diaphragmatic hernia with failure of fusion of pleuroperitoneal canal.

Micro Pathology N/A

Treatment Resuscitate and stabilize neonate, intubation, assisted ventilation followed by surgical repair.

Discussion The defect usually represents failure of the pleuroperitoneal canal to close completely (= FORAMEN OF BOCHDALEK), leading to protrusion of the abdominal viscera into the chest; it is usually located on the **left side.** Parasternal or retrosternal (= FORAMEN OF MORGAGNI) hernias are also congenital but usually do not produce symptoms so early and are located anteriorly (vs. Bochdalek's posterolateral location). **Pulmonary hypoplasia** is the most common cause of death in infants with diaphragmatic hernia.

. . . .

CONGENITAL DIAPHRAGMATIC HERNIA

ID/CC	A 37-year-old male is admitted to the ER following the development of marked **lightheadedness, sweaty palms, palpitations, and nausea.**
HPI	He has a history of duodenal ulcers that have been unresponsive to medical treatment, for which he underwent **surgery** two months ago (vagotomy and Billroth II anastomosis [**gastrojejunostomy**]).
PE	VS: **mild hypotension;** tachycardia; no fever. PE: abdominal exam discloses well-healed upper midline incision with no hematomas, dehiscence, or signs of infection; no peritoneal signs.
Labs	Hypoglycemia (50 mg/dL). CBC/Lytes: normal. LFTs, amylase normal.
Imaging	CT: no fluid collections in subphrenic, subhepatic, or pelvic spaces.
Gross Pathology	N/A
Micro Pathology	N/A
Treatment	Low-carbohydrate, high-protein, small, frequent, dry meals.
Discussion	A complication of duodenal surgery, dumping syndrome is due to the **rapid, unimpeded passage of high-osmolarity food to the jejunum,** with onset half an hour after meals (there is also a delayed type). Symptoms appear related to the development of **hypoglycemia.** The blood supply to the distal stomach-duodenum is derived from the gastroduodenal artery, a branch of the common hepatic artery. The pancreaticoduodenals are branches of the gastroduodenal, as is the right gastroepiploic, which courses through the greater curvature to join the left gastroepiploic artery, a branch of the splenic artery. The branches of the **celiac axis** are the **left gastric, splenic, and common hepatic arteries.**

· ·

DUMPING SYNDROME

ID/CC	A 57-year-old white male complains of deep, **burning retrosternal pain** (= HEARTBURN) that **worsens when he lies down.**
HPI	The patient is a heavy cigarette **smoker** and **alcohol** drinker. He also complains of **regurgitation** of sour material on and off for years. He has been **overweight** for the past 10 years and has recently experienced insomnia.
PE	Obese and moderately nervous; slight discomfort on palpation of epigastrium.
Labs	CBC may show anemia if ulcer is present.
Imaging	UGI: gastroesophageal junction and part of stomach protrude above diaphragm. EGD: may show esophageal inflammation.
Gross Pathology	N/A
Micro Pathology	Esophageal mucosa with variable degrees of inflammation.
Treatment	Weight loss, cessation of smoking, avoid lying down after meals, prokinetics, H$_2$ receptor blockers, proton pump inhibitors, surgery.
Discussion	Most hiatal hernias are **sliding** (the stomach herniates into the thorax together with the gastroesophageal junction, producing reflux), but they may also be **paraesophageal** (the gastroesophageal junction remains fixed below the diaphragm with no reflux; symptoms are due to pressure). Complications associated with paraesophageal hiatal hernias are strangulation, obstruction, incarceration, and hemorrhage. Chronic untreated **gastroesophageal reflux disease** secondary to a sliding hiatal hernia may lead to **Barrett's esophagus** (columnar metaplasia of the distal esophagus), which is associated with an increased risk of esophageal **adenocarcinoma.**

. . . .

HIATAL HERNIA

ID/CC	A 6-day-old **male** is brought to the emergency room with bilious vomiting, **abdominal distention, and failure to pass stools.**
HPI	The **full-term** baby **failed to pass meconium** in the first 24 hours after birth but did so immediately following a rectal exam.
PE	Abdomen **distended** and tympanic; loops of **intestine palpable;** increased anal tone; rectum empty; child passes foul-smelling stool following rectal exam.
Labs	N/A
Imaging	XR-Abdomen: **massively dilated colon** with gas and feces; rectal air normally visible in presacral area is absent on lateral erect view. BE: abrupt changes in caliber between ganglionic and aganglionic segments; failure to evacuate barium.
Gross Pathology	N/A
Micro Pathology	Rectal biopsy reveals abnormal development of Meissner's and Auerbach's plexuses with **aganglionosis in myenteric nerve and submucosa;** hypertrophy of nerve fibers in Meissner's plexus.
Treatment	Surgical excision of aganglionic segment and anastomosis to anal canal.
Discussion	Due to failure of migration of cells of embryonic neural crest to bowel wall of distal segments with absence of parasympathetic ganglion cells in the anal and rectosigmoid areas, leading to **functional** (not anatomic) **obstruction and colonic dilatation proximal to the affected segment.** May be associated with **Down's syndrome** and urinary anomalies. The presenting symptom may be acute **enterocolitis** with watery, foul-smelling diarrhea.

. . . .

HIRSCHSPRUNG'S DISEASE

ID/CC	A **3-week-old male** is brought to the pediatrician for **projectile, nonbilious vomiting** that began today, shortly after feeding.
HPI	He has been regurgitating food and has had occasional bouts of vomiting for one week. The child is the **first-born** son of a 28-year-old white female and is the product of a normal delivery. His **mother had hypertrophic pyloric stenosis** when she was born.
PE	Lethargic, moderately **dehydrated** baby; **low weight for age;** wrinkled, "old man" appearance; **visible peristalsis** from left upper quadrant toward right upper quadrant followed by projectile vomiting; hard, mobile, nontender, **olive-like mass felt in epigastrium** deep to right rectus muscle.
Labs	Lytes: hypokalemia; hyponatremia. ABGs: hypochloremic alkalosis (due to loss of gastric hydrochloric acid in vomitus).
Imaging	US: **pylorus muscle thickening** (target sign of pyloric stenosis); elongated pyloric canal; widened pylorus. UGI: pyloric wall thickening; elongated and narrowed pyloric channel (= STRING SIGN); vigorously peristaltic stomach with almost **no gastric emptying.**
Gross Pathology	Diffuse hypertrophy and hyperplasia of smooth muscle involving pyloric sphincter; muscular thickening extends proximally to antrum and ends where duodenum begins.
Micro Pathology	N/A
Treatment	Correction of fluid and electrolyte abnormalities, nasogastric tube decompression. Surgical relief of pyloric obstruction (= RAMSTEDT PYLOROMYOTOMY).
Discussion	Hypertrophic pyloric stenosis usually presents two weeks after birth, although it may not become apparent until several months of age.

. . . .

HYPERTROPHIC PYLORIC STENOSIS

ID/CC	A 73-year-old man is brought to the ER from his nursing home because of the sudden development of intense **abdominal pain**.
HPI	The patient states that the **pain is severe and even worse than his prior MI**. He has a history of similar but less severe **crampy abdominal pain after meals** (intestinal angina); he is a heavy **smoker**.
PE	On palpation, abdomen is only moderately tender and distended with no guarding; peristalsis not heard (**pain is out of proportion to clinical findings**); **rectal** exam shows **blood**.
Labs	CBC: marked leukocytosis (21,700) with neutrophilia; no anemia. Amylase moderately high; CK elevated.
Imaging	KUB: marked distention of bowel loops to splenic flexure; **gas in bowel wall**. BE: **thumbprinting** of bowel wall (due to submucosal hemorrhage and edema). Angio: vascular **occlusion by embolus**.
Gross Pathology	Surgical specimen reveals completely black and necrotic ileum and jejunum; multiple clots in **superior mesenteric artery (SMA)** branches.
Micro Pathology	N/A
Treatment	Immediate surgical intervention if massive; interventional angiographic thrombolysis if focal.
Discussion	Occlusive disease of the bowel may be due to thrombosis or emboli, giving rise to life-threatening intestinal infarction. The SMA is a direct branch of the aorta and supplies the right side of the colon, the appendix, and the jejunum and ileum (its branches are the middle colic, right colic, and ileocolic). The inferior mesenteric artery (IMA) also branches from the aorta and supplies the left colon, sigmoid, and upper rectum through its branches (left colic, sigmoid, and superior hemorrhoidal).

MESENTERIC ISCHEMIA

ID/CC	A 45-year-old female presents with **pain** and complains of **heaviness** and a **"tumor" in her abdomen**; she also has a **fever.**
HPI	Eight weeks ago she had been hospitalized for epigastric pain, nausea, and vomiting due to **acute pancreatitis.**
PE	VS: fever. PE: pallor; **epigastric mass** tender to palpation; mass not motile and seems to be deep-seated; no change of overlying skin; no peritoneal signs.
Labs	CBC: **elevated WBCs. Amylase and lipase elevated** (although not to extent of her first admission); AST and ALT slightly elevated; **bilirubin moderately increased.** UA: normal.
Imaging	CT/US: large cystlike fluid collection in close proximity to posterior wall of stomach, originating in pancreas.
Gross Pathology	Collection of enzyme-rich fluid around pancreas walled off by inflammatory adhesions of peritoneal surfaces, large bowel, and diaphragm; **no true capsule or epithelial lining** (= PSEUDOCYST).
Micro Pathology	N/A
Treatment	Imaging-guided intervention for placement of drainage catheter. Sometimes surgical drainage required.
Discussion	Pancreatic pseudocysts are a complication of pancreatitis that occur in about 3% of cases. The pancreas has a head, neck, body, and tail. The main pancreatic duct (= DUCT OF WIRSUNG) drains into the ampulla of Vater together with the common bile duct. The accessory duct (= DUCT OF SANTORINI) drains more proximally or into one of the above-mentioned ducts.

. . . .

PANCREATIC PSEUDOCYST

ID/CC	A **56-year-old male** bus driver is rushed to the emergency room with **generalized, excruciating abdominal pain that began in the epigastric area** after he ate a large meal; he also complains of nausea and vomiting.
HPI	He is a **heavy smoker.** For the past three years he has suffered from chronic, episodic, burning epigastric pain that **was diagnosed as a gastric ulcer** and treated with antacids and H$_2$ receptor blockers.
PE	VS: tachycardia (HR 110); mild hypotension (BP 100/60). PE: sweaty and in acute distress; marked, generalized abdominal tenderness, predominantly in epigastric area, with positive **rebound tenderness;** no peristalsis heard; **abdominal rigidity** (due to peritonitis).
Labs	CBC: neutrophilic leukocytosis. Slightly elevated amylase.
Imaging	XR-Abdomen: generalized small bowel loop dilatation. CXR: may show intraperitoneal subdiaphragmatic **free air.**
Gross Pathology	N/A
Micro Pathology	N/A
Treatment	Surgical removal of ulcer or closure of perforation in gastric wall, peritoneal lavage and drainage. Antibiotics for peritonitis. Treatment for *Helicobacter pylori* on discharge.
Discussion	Perforation is common in the gastric antrum or lesser curvature. The gastric contents may spill into the lesser or greater sac. The boundaries of the lesser sac are the hepatoduodenal ligament, caudate lobe of liver, duodenum, and inferior vena cava. The lesser sac communicates with the peritoneal cavity via the foramen of Winslow. Anterior duodenal ulcers can cause perforation, while posterior duodenal ulcers are associated with hemorrhage secondary to ulcer erosion into the gastroduodenal artery.

· ·

PERFORATED GASTRIC ULCER

ID/CC	A 47-year-old male is brought by ambulance to the emergency room **vomiting copious amounts of blood** (= MASSIVE HEMATEMESIS).
HPI	He has a history of **heavy alcoholism**; he has gotten drunk at least three times a week for many years.
PE	VS: **tachycardia** (HR 103); **hypotension** (BP 90/40) (due to hypovolemia); no fever. PE: marked **pallor**; thin, wasted, delirious man with strong **alcohol smell on breath**; pupils reactive and equal; enlargement of parotid glands; no focal neurologic signs; abdomen enlarged due to **ascitic fluid**; **spider angiomas** over abdominal skin; **palmar erythema**.
Labs	CBC: low hemoglobin (7.3 mg/dL); leukocytosis. **Increased ALT and AST**; mild hyperbilirubinemia.
Imaging	Esophagoscopy: active bleeding of markedly dilated and tortuous submucosal veins (= BLEEDING VARICES).
Gross Pathology	Alcoholic hepatitis gives rise to **fibrosis** (= CIRRHOSIS) of the liver, which **increases portal vein resistance**. With the development of **portal hypertension** (>10 mmHg), there are portal-systemic anastomoses formed such as the **left gastric-azygous (esophageal varices)**, the **superior-middle and inferior rectal veins (hemorrhoids)**, the **paraumbilical-inferior gastric (navel caput medusae)**, and the retroperitoneal-renal vein system.
Micro Pathology	N/A
Treatment	Sclerotherapy. In emergency bleeding, balloon tamponade, endoscopic cauterization, ligation, IV vasopressin, surgery. Consider splenorenal or transhepatic portal-systemic shunt.
Discussion	The portal vein is formed by the joining of the mesenteric vein and the splenic vein; tributaries include the left and right gastric veins. On occasion the inferior mesenteric vein drains into the superior mesenteric vein rather than into the splenic vein. **FIRST AID** p.230

ID/CC	A 58-year-old obese man presents for an evaluation of a **"lump" in the anal area** of three days' duration, causing **acute, constant pain** that increases during defecation.
HPI	He is a smoker with a **chronic cough** and is **overweight.** He also suffers from **prostatic hyperplasia** that forces him to strain in order to initiate micturition.
PE	Patient walks very slowly with both legs apart and sits down in chair sideways; external rectal exam reveals presence of a **rounded,** 3-cm, **purple mass** in the anal verge that is **tense and extremely painful to the touch;** internal digital rectal exam impossible due to acute pain; mass localized to outer anal region.
Labs	CBC: slight leukocytosis.
Imaging	N/A
Gross Pathology	Dilated, engorged vein with clot.
Micro Pathology	Acute inflammatory neutrophilic infiltrate.
Treatment	Acute thrombosis will subside spontaneously in most cases with **sitz baths,** anti-inflammatories, local steroids, and laxatives. If recurrent, surgical resection is warranted. If acutely painful or if conservative treatment fails, excision with local anesthesia may be done.
Discussion	External hemorrhoids are dilatations of the anal veins from the inferior hemorrhoidal plexus, which drains into the internal pudendal veins. Internal hemorrhoids lie above the mucocutaneous junction (pectinate line) and belong to the superior hemorrhoidal plexus, which drains into the portal vein through the inferior mesenteric vein. Internal hemorrhoids are painless (visceral innervation) and are covered by mucosa. External hemorrhoids are painful (somatic innervation) and are covered by skin.

. . . .

THROMBOSED EXTERNAL HEMORRHOIDS

ID/CC   A **newborn** male baby presents with inability to accept food; **he chokes, coughs, and vomits with each attempt to feed him.**

HPI   Prenatal ultrasound showed **excess amniotic fluid** (= POLYHYDRAMNIOS) **and no fluid in stomach.**

PE   Full-term baby; **excessive salivation;** abdomen distended and tympanitic; **catheter cannot be passed into stomach;** chest exam normal.

Labs   N/A

Imaging   CXR: coiled feeding catheter in upper esophageal pouch; **gastric air bubble present.**

Gross Pathology   The most common type of tracheoesophageal fistula is that associated with a blind proximal esophageal pouch and a distal esophageal pouch that communicates via a fistula with the lower trachea.

Micro Pathology   N/A

Treatment   Treat aspiration pneumonia. Keep esophageal pouch empty by constant suction. Surgical repair as early as possible.

Discussion   Anomalies are due to defective differentiation of primitive foregut into the trachea and esophagus, defective growth of endodermal cells leading to atresia, and incomplete fusion of the lateral walls of the foregut during separation of the trachea from the foregut. There may be esophageal atresia without tracheoesophageal fistula and tracheoesophageal fistula without esophageal atresia. Maternal **polyhydramnios** is associated with **esophageal/duodenal atresia and anencephaly** (the fetus cannot swallow amniotic fluid), while maternal **oligohydramnios** is associated with **bilateral renal agenesis and posterior urethral valves** (fetal urine is absent or obstructed).

. . . .

TRACHEOESOPHAGEAL FISTULA

ID/CC	A 12-week-old male infant is brought to the hospital for evaluation of **recurrent oral thrush and URIs.**
HPI	The child had **seizures** (due to hypocalcemia) shortly after birth. His mother is an IV drug user.
PE	Full-term infant; jittery; increased muscle tone; oral thrush; midfacial hypoplasia.
Labs	**Hypocalcemia; T-cell count markedly low.**
Imaging	CXR: **absent thymic shadow.**
Gross Pathology	N/A
Micro Pathology	N/A
Treatment	Transplant of an immature fetal thymus; treat opportunistic infections; irradiation of blood products.
Discussion	Due to an embryologic defect characterized by lack of development of the **third and fourth pharyngeal pouches.** Associated with **lack of thymus** and thus no cell-mediated immunity (hence recurrent viral and fungal infections) as well as with congenital heart defects. The associated **lack of parathyroid glands** results in hypocalcemia, leading to tetany and/or convulsions. **FIRST AID** p.209

. . . .

DIGEORGE'S SYNDROME (THYMIC APLASIA)

ID/CC	A 65-year-old Vietnam veteran complains of progressively worsening **hoarseness and persistent cough.**
HPI	He has **smoked** one pack of cigarettes each day for 45 years. He is currently being treated for emphysema.
PE	Supraclavicular nodes **hard and enlarged** (cancer has metastasized to these sentinel nodes); lung fields filled with disseminated crackles and rales; chest barrel-shaped (underlying emphysema); clubbing of fingers on both hands; **patches of velvety hyperpigmentation** (= ACANTHOSIS NIGRICANS) of both lower legs; purpuric spots seen on chest and arms.
Labs	CBC: secondary **polycythemia** and leukocytosis.
Imaging	CXR: 4-cm **mass** at left **lung hilum** with thickening of paratracheal stripes and hilar fullness.
Gross Pathology	N/A
Micro Pathology	Sputum cytology and lymph node biopsy show small-cell carcinoma.
Treatment	Depends on stage; surgery, radiotherapy, chemotherapy, neoadjuvant and immunotherapy. Overall, around 9% five-year survival rate.
Discussion	Lymphatic drainage of the lung goes to the bronchopulmonary nodes (at hila of lungs), tracheobronchial nodes (superior and inferior sets around trachea and main bronchus), paratracheal nodes, and bronchomediastinal lymph trunk (which form at the junction of the tracheobronchial nodes, the anterior mediastinal nodes, and the parasternal nodes). Only when pleural adhesions are present do the axillary nodes drain the lung. The lower lobe of the left lung drains to the right tracheobronchial nodes. Small cell (= OAT CELL) carcinomas metastasize early to lymph nodes.

LYMPHATIC METASTASIS OF LUNG CANCER

ID/CC	A 63-year-old woman who is a heavy **smoker** comes to the emergency room with severe **swelling on the right side of the neck,** arm, and face (compression of superior vena cava) together with severe pain in the right arm.
HPI	The swelling has gotten progressively worse over the past several months. In addition, her voice has become **hoarse** over the same period of time (the recurrent laryngeal nerve is paralyzed due to compression by the tumor).
PE	**Reduced radial pulse** on right side (arterial flow is blocked by tumor); **engorgement of right jugular vein** (venous return is blocked due to impingement by apical tumor); slight **ptosis** (drooping) of right eyelid, **miosis** (contraction of pupil), and **anhidrosis** (lack of sweating/lacrimation) (= HORNER'S SYNDROME; due to sympathetic chain compression); **wasting of first dorsal interosseous muscle** of right hand (supplied by T1); pain and muscle atrophy (involvement of brachial plexus) of right arm.
Labs	CBC: anemia. Lytes: normal. BUN and creatinine normal; LFTs normal.
Imaging	CXR: lung tumor in right apex, destroying first rib.
Gross Pathology	Apical lung tumor has invaded cervical sympathetic plexus and vena cava.
Micro Pathology	Bronchial washings and Papanicolaou of sputum show squamous cell carcinoma.
Treatment	Depending on stage; surgery, radiotherapy, chemotherapy, neoadjuvant and immunotherapy.
Discussion	Produced by **any tumor in close proximity to the thoracic inlet** and by the consequent compression of the brachial plexus and both the venous return and arterial flow.

. . . .

PANCOAST'S SYNDROME

ID/CC   A 66-year-old diabetic high-school biology teacher goes to a surgeon for an evaluation of a small **nodule on his lower lip.**

HPI   The nodule has been there for over a year, and he wonders if it might be related to his habit of **chewing tobacco** while watching baseball games after school.

PE   Indurated, ill-defined, violaceous, nonmotile mass felt in lower lip; enlargement of **submental lymph nodes** (which drain the medial portion of the lower lip) and **submandibular lymph nodes** (which drain the lateral portion of the lower lip); enlargement of **deep cervical lymph nodes** near omohyoid muscle.

Labs   CBC: mild anemia.

Imaging   CT-Neck: demonstrates involved lymph nodes.

Gross Pathology   Surgical pathology specimen consists of a wedge of lower lip with an ulcerated lesion and rolled edges.

Micro Pathology   Lymph node biopsy shows metastatic squamous cell carcinoma.

Treatment   Surgical removal of nodule on lip and radical neck dissection (removal of submandibular gland [adhered to affected submandibular nodes], sternocleidomastoid muscle, omohyoid, accessory nerve, and internal jugular vein).

Discussion   Important regions in the anatomy of this disease include the submandibular triangle (between the two anterior bellies of the digastric containing the submandibular lymph nodes); muscular triangle (omohyoid, sternocleidomastoid, median line, containing some deep cervical lymph nodes); carotid triangle (sternocleidomastoid, omohyoid, posterior belly of digastric, containing internal jugular vein and deep cervical nodes); and posterior cervical triangle (trapezius, sternocleidomastoid, omohyoid, containing deep cervical nodes near accessory nerve).

. . . .

SQUAMOUS CELL CARCINOMA OF THE LIP

ID/CC	A 59-year-old female comes to her family physician because of left-sided **hearing loss,** numbness over the left half of her face, and **unsteadiness of gait** of about one month's duration.
HPI	She also complains of intermittent **vertigo** and **ringing in the ear** (= TINNITUS). She has no history of earache, ear discharge, or eruption over the pinna.
PE	Patient falls to left when standing with eyes closed (= POSITIVE ROMBERG'S SIGN); fundus normal; gait wide and ataxic; when tuning fork is placed in midline of skull, patient reports that she hears best on right side (Weber test lateralizes to the right, i.e., the normal side, because the left suffers a sensorineural loss); caloric testing shows **left canal paresis.**
Labs	Audiometry shows **sensorineural hearing loss** on left side that is more pronounced with **high frequencies.**
Imaging	CT/MR: contrast-enhanced **cerebellopontine (CP) angle** mass extending into internal auditory canal.
Gross Pathology	Slow-growing schwannoma of **eighth nerve.**
Micro Pathology	N/A
Treatment	Surgical removal.
Discussion	Acoustic schwannomas are Schwann cell–derived neoplasms that comprise about three-fourths of CP-angle neoplasms. They arise mainly from the vestibular division of CN VIII. Other CP-angle masses include meningioma, arachnoid cyst, and epidermoid tumor. Bilateral acoustic schwannomas are seen in **type 2 neurofibromatosis** (NF-2).

. . . .

ACOUSTIC SCHWANNOMA

ID/CC	A 42-year-old man presents to his family doctor complaining of **pain and stiffness** on **one side of his neck** that precludes normal movements.
HPI	The patient is obese and sedentary and never exercises. His pain started after he went outside and shoveled the first snow without warming up (sudden, vigorous physical exercise).
PE	**Head tilted** to one side; patient cannot straighten head without considerable pain; accompanied by considerable **muscle spasm** in left side of neck.
Labs	N/A
Imaging	XR-Cervical Spine: no fracture or subluxation.
Gross Pathology	N/A
Micro Pathology	N/A
Treatment	Muscle relaxants, NSAIDs, local heat, massage.
Discussion	Caused by **acute spasm of neck muscles** due to inflammatory changes because of undue straining; the muscles that are usually involved are the trapezius, supraspinatus, rhomboid, splenius capitis, levator scapula, scalenus medius, splenius cervicis, and, in severe cases, transverse ligament, allowing subluxation of one vertebra on another. Torticollis may also be congenital due to unilateral fibrosis of the sternocleidomastoid muscle.

. . . .

ACUTE TORTICOLLIS

ID/CC   A 7-year-old girl is brought by her parents to the pediatric emergency department because of a **severe headache** that does not respond to treatment with analgesics.

HPI   Her father is concerned about "weird movements" of the girl's eyes (= NYSTAGMUS).

PE   Well developed and nourished but confused with **ataxic gait**; chest and abdominal exams unremarkable; funduscopic exam reveals left **papilledema** (due to increased intracranial pressure); nystagmus also noted; patient experienced projectile vomiting during examination (also due to increased intracranial pressure).

Labs   N/A

Imaging   MR/CT: **cystic posterior fossa tumor**.

Gross Pathology   Grayish cystic mass; zones of necrosis, hemorrhage, and calcification; cerebral edema.

Micro Pathology   N/A

Treatment   Surgery, chemotherapy, radiotherapy.

Discussion   Astrocytomas are slow-growing **malignant tumors that originate from neuroectodermal neuroglia.** In children they are usually located in the cerebellum (posterior fossa), whereas in adults they are located in the cerebrum. Often cystic in children, their growth may cause increased intracranial pressure, seizures, and hydrocephalus. The posterior cranial fossa is limited anteriorly by the dorsum sella, laterally by the parietal bones, and posteriorly by the occipital bone; it contains the foramen magnum for the spinal cord as well as the jugular foramen and the internal acoustic meatus.

. . . .

ASTROCYTOMA

ID/CC	A 45-year-old diabetic male comes to the emergency room fearing that he has had a stroke; he complains of **inability to move the right side of his face** and **cannot blink his right eye or seal his lips** (lesion of CN VII - facial nerve).
HPI	The previous night, his wife noticed that he was **sleeping with his right eye open** and that the right side of his face was drooping. That morning, the patient could not stop **drooling** on the right side of his mouth. He has not closely monitored his blood sugar for several months.
PE	Right-sided facial muscles flaccid; under forced closure of right eye, eyeball rotates upward (= BELL'S PHENOMENON); cognitive function normal; when patient is asked to smile, right side of mouth remains flaccid.
Labs	CBC/Lytes: normal. Hyperglycemia; LFTs normal.
Imaging	N/A
Gross Pathology	Inflammation of CN VII in the vicinity of the stylomastoid foramen.
Micro Pathology	N/A
Treatment	Steroids, artificial tears or eye covering.
Discussion	Bell's palsy is seen as a complication of diabetes, AIDS, Lyme disease, tumors, and sarcoidosis but is most commonly idiopathic. Involvement of the entire half of face demonstrates lower motor neuron (CN VII) pathology; involvement of only the lower half of the face suggests upper motor neuron pathology (upper motor neurons have cross-innervation to the forehead). It is self-limited in most cases (typically resolves in 6–12 months), with only a small percentage of cases resulting in permanent disfigurement. Thought to represent a viral cranial polyneuropathy. **FIRST AID** p.117

· ·

BELL'S PALSY

ID/CC	A 45-year-old man is brought by ambulance to the emergency department of the local community hospital complaining of **inability to move his left leg.**
HPI	He was **stabbed in the back** two hours ago while defending his wife from a mugger.
PE	Moderate bleeding; stab wound at level of the posterior cervical spinous prominence (C7) on left side; **loss of position sense** of left leg; **weakness** of finger flexion; extension of left finger; **inability to sense vibration** of tuning fork along left **lower limb; loss of pain and temperature sense in contralateral lower limb.**
Labs	N/A
Imaging	MR: hematoma at level of C7–T1 in left half of spinal cord.
Gross Pathology	Hemisection and compression of spinal cord at level of C7.
Micro Pathology	N/A
Treatment	Surgical removal of hematoma.
Discussion	Consists of ipsilateral upper and lower motor paralysis below the level of the injury (corticospinal tract), ipsilateral cutaneous anesthesia, ipsilateral loss of vibrations and proprioception (dorsal column), and contralateral loss of pain and temperature sense below the level of the lesion (spinothalamic tract). Brown–Séquard syndrome is usually due to a penetrating injury to the spine, resulting in **functional hemisection of the spinal cord.** **FIRST AID** p.115

. . . .

BROWN–SÉQUARD SYNDROME

ID/CC	A 45-year-old boy-scout instructor returns from a two-week camping trip with a high fever, a severe headache, and a **pus-filled boil on his right cheek** that appeared after he cut himself on a tree branch.
HPI	The patient has a long history of **diabetes mellitus** that has been treated with insulin. He also complains of intermittent vomiting, nausea, and episodes of delirium. His headache is particularly severe on the right side.
PE	VS: fever. PE: neck muscles stiff; **right cheek swollen and red** with area of purulent discharge; **right side of nose hard and swollen;** right eye very painful and protrudes (= EXOPHTHALMOS); right eyelid swollen with black discoloration; loss of function of right extraocular eye muscles; tingling and burning (= PARESTHESIA) of right upper quadrant of face.
Labs	*S. aureus* on blood and wound pus culture.
Imaging	CT/MR: lack of cavernous sinus enhancement; clot in cavernous sinus.
Gross Pathology	Septic venous thrombosis blocking tributaries of orbit; eyelid edema and discoloration.
Micro Pathology	N/A
Treatment	Aggressive course of IV antibiotics.
Discussion	The infection began at the injury site and progressed along the facial vein and superior ophthalmic vein to one of the paired cavernous sinuses, one on each side of the sella turcica; lack of valves in the veins of the face facilitated migration of infectious thrombosis throughout the face. Pulmonary septic thrombi and **meningitis** are common complications.

· ·

CAVERNOUS SINUS THROMBOSIS

ID/CC	A 34-year-old construction worker is brought to a clinic after a flying piece of shrapnel **cut his leg** just below the **lateral surface of the head of the fibula;** the patient also complains of **numbness and tingling along the dorsum of his foot** and the lateral surface of his leg.
HPI	He complains of loss of stability when walking and **inability to dorsiflex his foot** (= FOOT DROP). He must raise his injured leg higher than normal during walking to prevent his toes from hitting the ground, and his **foot slaps against the ground** when walking (steppage gait due to unopposed action of plantar flexors; muscles innervated by peroneal nerve are paralyzed: extensor digitorum longus, tibialis anterior, extensor hallucis longus).
PE	Wound in proximity of head of fibula (area where the common fibular nerve is most superficial as it wraps around the head of the fibula); **diminished cutaneous sensation** over anterolateral aspect of leg and dorsum of foot (cutaneous branches of the superficial fibular nerve supply this area); inability to extend toes (paralysis of tibialis anterior, peroneus longus and brevis).
Labs	N/A
Imaging	XR-Leg: fracture of head of fibula.
Gross Pathology	N/A
Micro Pathology	N/A
Treatment	Spring-loaded brace of foot to prevent foot drop during walking and to provide additional stability.
Discussion	The common peroneal nerve is frequently injured due to trauma of the upper leg (e.g., broken fibula) owing to its superficial location around the lateral surface of the head of the fibula. **FIRST AID** p.96

. . . .

COMMON PERONEAL NERVE DAMAGE

ID/CC	The parents of a 9-year-old boy are called into his teacher's office to talk about academic problems the child has been having; the teacher suspects that the child **cannot hear properly.**
HPI	Since infancy, he has had **recurrent ear infections** with discharge.
PE	Otoscopic exam shows **perforation of right tympanic membrane;** when tuning fork is placed in midline of skull, patient reports he hears best on right side (= WEBER TEST LATERALIZES TO RIGHT); bone conduction greater than air conduction in right ear (= POSITIVE RIGHT RINNE TEST).
Labs	N/A
Imaging	XR-Cranium and Sinus Cavities: within normal limits.
Gross Pathology	N/A
Micro Pathology	N/A
Treatment	Myringoplasty is definitive treatment.
Discussion	This is a case of conductive (not nerve-associated) deafness in the right ear secondary to a tympanic membrane perforation. Normally, air conduction is greater than bone conduction (i.e., it gives a negative Rinne test), whereas the reverse is the case in conductive deafness. The Weber test lateralizes to the affected ear in cases of conductive deafness and to the normal ear in cases of sensorineural hearing loss. Interpretation of the two tests together can identify the type of hearing loss.

CONDUCTIVE DEAFNESS

ID/CC	A 9-year-old boy complains of **pain** in his left **elbow** that began **after he fell** off his bicycle, hitting the ground elbow first.
HPI	He also has **numbness on the medial side of his hand** (due to damage of the ulnar nerve at the medial epicondyle–olecranon groove).
PE	Elbow skin shows dermal abrasions and soft tissue edema with tenderness on palpation; **inability to abduct fingers; poor grasp of fourth and fifth digits** (due to ulnar nerve damage; all but five of the interosseous muscles are innervated the ulnar nerve).
Labs	N/A
Imaging	XR-Elbow: separation of epiphysis of medial epicondyle (children under 16 have unfused epiphyseal plate).
Gross Pathology	N/A
Micro Pathology	N/A
Treatment	Isolation of elbow after reunion of fractured ends; physical therapy for hand movement (crushed nerve will regenerate).
Discussion	Of all injuries to long bones during childhood, approximately 15%–20% involve the growth plate. A growing bone subjected to a shearing force may cause the epiphysis to separate from the growth plate, producing signs and symptoms of a fracture. Improper healing of the growth plate may result in length or bowing deformity.

. . . .

EPIPHYSEAL SEPARATION WITH ULNAR NERVE PALSY

ID/CC	A 45-year-old female is scheduled to undergo a left **parotidectomy** due to a tumor.
HPI	Upon awakening from anesthesia after removal of the gland, she was asked to smile but could not do so properly.
PE	Surgical wound covered by sterile gauze with no apparent bleeding; small drain in place; when patient is asked to wrinkle forehead, patient's left side would not wrinkle (affected side); left corner of mouth does not rise up as right one does when patient is asked to smile; patient **cannot raise her left eyebrow** or lower it when asked to frown; **mouth does not lift on left side** when patient is asked to show teeth.
Labs	N/A
Imaging	N/A
Gross Pathology	Surgical specimen consists of a malignant adenocarcinoma of parotid gland; a 0.5-cm-long portion of facial nerve was found resected, enmeshed in carcinomatous lobules.
Micro Pathology	N/A
Treatment	Facial nerve can be sutured or grafted with sural nerve if lesion is voluntary or involuntary but cannot be overlooked; otherwise, physiotherapy.
Discussion	Muscles innervated by the facial nerve include most of the muscles that move the face and scalp, including the buccinator and the posterior belly of the digastric. The facial nerve gives taste sensation to the anterior two-thirds of the tongue. The parotid gland consists of two lobules, the anterior and posterior; the facial nerve and its branches (cervical, buccal, zygomatic, temporal, and mandibular) course between the two after exiting through the stylomastoid foramen.

FACIAL NERVE INJURY

ID/CC	A 26-year-old professional cyclist visits a sports medicine doctor complaining of **weakness in the right leg as well as lack of sensation** (= ANESTHESIA) **in the anterior area of his thigh.**
HPI	Eight weeks ago, he fell from his bicycle during a race and suffered a **pelvic fracture.** He was treated through use of a sling and bed rest and is now trying to begin rehabilitation with crutches.
PE	Patient has significant **weakness with extension of his right knee;** when patient is asked to stand, there is obvious **instability** despite x-ray consolidation of fracture, and **walking is impossible** (due to weak hip flexion).
Labs	N/A
Imaging	XR: healed pelvic fracture.
Gross Pathology	N/A
Micro Pathology	N/A
Treatment	Physiotherapy, surgical exploration and repair in selected cases.
Discussion	The quadriceps femoris muscle is innervated by the femoral nerve. When there is a nerve injury, as is the case in penetrating wounds or pelvic fractures, the patient is unable to extend the knee and there is anesthesia in the anterior area of the lower extremity from the thigh to the foot.

FIRST AID p.96

FEMORAL NERVE PALSY

ID/CC	A 42-year-old man complains of **lower back pain** that began after he **lifted heavy objects** while helping his son move out of the family home.
HPI	He is overweight and has not had any regular exercise for the past 10 years. The pain is **aggravated by movement, coughing, and sneezing;** it **radiates down his buttocks, thigh, and posterior calf.**
PE	**Sensory loss over dorsal aspect of foot** and lateral aspect of leg (**L5 dermatome**); **weakness of dorsiflexors of foot**; on palpation, left sciatic notch is tender; positive Lasègue's sign (straight leg-raising test); deep tendon reflexes normal.
Labs	N/A
Imaging	MR-Lumbar Spine: focal herniated disk centrally at L4–L5 touching L5 root.
Gross Pathology	N/A
Micro Pathology	N/A
Treatment	Physiotherapy; bedrest; acupuncture; chiropractic therapy if symptoms are mild. Consider surgical laminectomy or laminotomy if pain becomes progressive or neurologic deficits are present.
Discussion	Herniation of the intervertebral disk (most commonly occurs at L5–S1) can lead to impingement on spinal nerve roots. Region of anesthesia can be used to deduce the specific level of nerve impingement. Central disk herniation at the L4–L5 level will cause compression of the L5 nerve root, while peripheral disk herniation at L4–L5 can affect the L4 nerve root. Signs and symptoms of L4 nerve root compression include an abnormal patellar deep tendon reflex, numbness over the medial aspect of the leg, and weakness of the tibialis anterior muscle (foot dorsiflexion). L5 nerve root compression causes numbness over the lateral aspect of the leg and weakness of the extensor hallucis longus.

. . . .

HERNIATION OF INTERVERTEBRAL DISK

ID/CC	A 28-year-old man with **tuberculous adenitis of the neck** is being treated with a multiple drug regimen; two days ago he developed "**strange movements**" of the tongue and cannot stick his tongue out normally.
HPI	He has been HIV positive for two years.
PE	When patient is asked to **protrude his tongue, it deviates to the left** (deviation of the tongue to the affected side is caused by the unopposed action of the contralateral genioglossus muscle, which is normally innervated); left side of tongue **atrophied and flaccid** with **fasciculations.**
Labs	HIV positive; PPD (tuberculin skin test) positive.
Imaging	N/A
Gross Pathology	N/A
Micro Pathology	N/A
Treatment	Treat cause (TB, tumor, etc.); physiotherapy.
Discussion	A flaccid paralysis accompanied by atrophy of the tongue denotes a lower motor neuron lesion of the hypoglossal nerve. Causes include parotid and carotid body tumors, tuberculous adenitis, and metastatic neck tumors.

ID/CC	A newborn child is noted on his first complete physical examination to have his **right hand in a "claw" position.**
HPI	He is the son of an 18-year-old female who was attended at home by a midwife; he was born with **shoulder dystocia.**
PE	In addition to claw hand, examination discloses pupillary constriction (= MIOSIS) with right-drooping eyelid (= PTOSIS) and lack of sweating (= ANHIDROSIS).
Labs	Neonatal enzyme deficiency screening and basic lab work normal.
Imaging	XR: no fracture of clavicle.
Gross Pathology	N/A
Micro Pathology	N/A
Treatment	Physiotherapy, nerve repair in selected cases.
Discussion	Klumpke's is an **abduction** injury affecting the lower brachial plexus, whereas Erb's is an **adduction** injury affecting the upper brachial plexus. Paralysis of the lower brachial plexus primarily affects the C7, C8, and T1 roots and produces paralysis of the muscles innervated by them (e.g., ulnar nerve, claw hand). Concomitant damage to the sympathetic fibers of T1 may produce **Horner's syndrome** (= MIOSIS, PTOSIS, ANHIDROSIS).

· ·

KLUMPKE'S PALSY

ID/CC	An 8-year-old male presents with progressive **dysarthria, dysphagia,** and **weakness** of the right side of his body of two months' duration.
HPI	The child has no history of fever, vaccinations, exanthem, dog bites, or travel outside the United States.
PE	Fundus normal; **atrophy of left side of tongue; joint position and vibration sense reduced on right side;** spastic right-sided hemiparesis; **face spared;** deep tendon reflexes brisk on right side; **extensor plantar response noted on right side;** no cerebellar signs present.
Labs	N/A
Imaging	CT-Head: left medial medullary enhancing mass with edema.
Gross Pathology	N/A
Micro Pathology	Infiltrating glioma.
Treatment	Inoperable tumor because of surgically inaccessible site; irradiation is the primary form of treatment.
Discussion	Midline structures are involved, including CN XII (= HYPOGLOSSAL), pyramidal motor tracts, medial lemniscus (proprioceptive sensation), and medial longitudinal fasciculus (connects various CN nuclei). The **hallmark of brainstem lesions is ipsilateral cranial nerve palsy with contralateral hemiplegia.** Lesions of the midbrain produce partial ophthalmoplegia and contralateral hemiplegia. Lesions of the pons produce ipsilateral paralysis of conjugate gaze or internuclear ophthalmoplegia, contralateral hemiplegia, and loss of position and vibratory senses. Lesions of the medulla produce paralysis of the tongue on the same side and paralysis of the contralateral limbs; the face is spared.

. . . .

MEDIAL MEDULLARY SYNDROME

ID/CC	A 9-month-old infant is brought to his family physician because his parents are worried that the child's **head appears too large.**
HPI	The mother had an apparently uneventful pregnancy and delivery. At birth the child's body weight and head circumference were at the 65th percentile (normal).
PE	Lethargic and irritable; anterior fontanelle bulging; when pressed slightly, it immediately pops back (increased intracranial pressure); head circumference enlarged (an infant's head enlarges with increased intracranial pressure, since fusion of cranial sutures is incomplete).
Labs	RPR in mother negative (for syphilis).
Imaging	MR-Head: **dilated lateral ventricles;** dilated third ventricles; **stenosis of cerebral aqueduct** (noncommunicating hydrocephalus).
Gross Pathology	Congenital obstruction of aqueduct of Sylvius due to arachnoiditis, with marked ventricular dilatation and atrophy of cerebral tissue (most common site of congenital obstruction is at the aqueduct).
Micro Pathology	N/A
Treatment	Surgical insertion of shunt either from lateral ventricle to inferior vena cava or directly from third ventricle to subarachnoid space.
Discussion	Almost half of infants with hydrocephalus will have **Arnold–Chiari syndrome** (hydrocephalus, syringomyelia, platybasia, and myelomeningocele). Other causes include infections (TORCH), but can also be idiopathic, as in this case. Noncommunicating (obstructive) hydrocephalus results from obstruction to CSF flow within the ventricles, causing dilation of the ventricles upstream of the block. **Communicating** (nonobstructive) **hydrocephalus results from failure of CSF reabsorption** in the subarachnoid space.

. . . .

OBSTRUCTIVE HYDROCEPHALUS

ID/CC	A **50-year-old obese man** comes to see his physician at the urging of his wife; she states that her husband **sleeps restlessly and has headaches upon awakening** (due to inability to breathe while sleeping).
HPI	He is a heavy **smoker.** His wife complains that his **loud snoring** is keeping her up at night. The patient also feels very tired during the day despite the fact that he gets 10 hours of sleep each night.
PE	VS: **hypertension** (BP 160/100). PE: patient **markedly obese** with bull's-neck appearance; **tongue large; nasal septum deviated** to left; heart sounds reveal **arrhythmic rate** (prolonged anoxia); lungs hypoventilated; pitting edema in legs.
Labs	CBC: polycythemia (compensatory effort for hypoxia). LFTs and thyroid function tests normal. ECG: premature ventricular contractions.
Imaging	Polysomnography: cyclic apneic episodes.
Gross Pathology	N/A
Micro Pathology	N/A
Treatment	Losing weight is most important measure. Adjuvant therapy is protriptyline, nasal positive pressure mask, repair of septum, oxygen.
Discussion	The syndrome is seen in middle-aged males who are usually morbidly obese, smokers, and hypertensive. It is due to a number of causes, mainly obesity, pharyngeal malformations, drugs, and alcohol. Patients present cyclical periods of hypoventilation and apnea sometimes lasting minutes, which cause anoxia, arrhythmias, and lack of normal sleep. Results in poor physical well-being during the day, mood changes, and work and family problems.

. . . .

OBSTRUCTIVE SLEEP APNEA

ID/CC	A 65-year-old **male** visits his family medicine clinic because of slowing of voluntary movements (= BRADYKINESIA), **unstable gait,** and **muscular rigidity.**
HPI	He also complains of **tremor at rest** that worsens when his grandchildren come to the house and make a lot of noise (emotional tension).
PE	Seborrheic dermatitis on scalp; infrequent blinking, with **masklike** (flat) **facies;** cardiopulmonary and abdominal examination within normal limits; **muscle rigidity** with passive movements (= COGWHEEL RIGIDITY); **pill-rolling movement** of hands and fingers (= RESTING TREMOR) (vs. cerebellar tremor - intention tremor).
Labs	Serum copper levels normal (vs. Wilson's disease).
Imaging	N/A
Gross Pathology	Depigmentation of substantia nigra.
Micro Pathology	**Decreased dopamine concentration** in substantia nigra, locus ceruleus, and striatum (= LENTICULAR AND CAUDATE NUCLEI); intracytoplasmic inclusion bodies (= LEWY BODIES) in substantia nigra.
Treatment	Bromocriptine (dopamine agonist), anticholinergics, levodopa (to produce dopamine), selegiline (selective MAO-B enzyme inhibitor).
Discussion	Also called paralysis agitans, it is an idiopathic disorder with a male predominance; characterized by decreased dopamine due to basal ganglia degeneration, mainly in the substantia nigra, with a resultant relative excess of acetylcholine. Since dopamine cannot cross the blood–brain barrier, levodopa, a dopamine precursor, is given; levodopa crosses the blood–brain barrier and is converted to dopamine in the brain.

. . . .

PARKINSON'S DISEASE

ID/CC   A 2-week-old child is referred to the pediatric surgeon because of a fleshy **mass in his lower back** that has been **present since birth.**

HPI   He is the second-born child of a 38-year-old woman who did not seek prenatal care until the time of delivery (i.e., she took **no folic acid** during pregnancy).

PE   Head diameter normal for age; patient found to have pes cavus and arched legs; deep tendon reflexes **hyporeflexic;** rounded, **large mass that transilluminates** partially seen in lumbosacral area.

Labs   CBC/Lytes: normal.

Imaging   XR: lumbar spine defect in L5 neural arch; lamina unfused; widened canal; soft tissue mass seen on lateral film. MR: mass communicates with spinal canal.

Gross Pathology   Failure of fusion of neuropore; spinal cord (neuroectoderm derived) and meninges (mesoderm derived) are outpouched; skin (ectoderm), muscle (myotome), and bone (sclerotome) have not developed over surface properly; ependymal, mantle, and marginal layers of primitive spinal cord have not developed.

Micro Pathology   N/A

Treatment   Early surgery.

Discussion   The **most common** developmental defect of the central nervous system; **involves incomplete fusion of the dorsal vertebral arches;** often associated with **hydrocephalus.** There are several degrees, from spina bifida occulta, where no defect is seen and the skin is intact, to meningocele and myelomeningocele, where leptomeningeal and neural tissue may protrude through a defect in the dura mater, bone, and skin, usually in the lumbosacral area. Lack of folic acid during pregnancy is associated with spina bifida. Associated with elevated maternal serum alpha-fetoprotein.
FIRST AID p.220

. . . .

SPINA BIFIDA WITH MYELOMENINGOCELE

ID/CC	A 46-year-old male is brought to the emergency room because he **suddenly** developed a **severe headache** (he states it is his worst ever) and **drooping of the left eyelid** (= PTOSIS).
HPI	His medical history is unremarkable; he is taking no medications and has seen his doctor regularly for annual checkups.
PE	**Dilated left pupil** (= MYDRIASIS); **outward deviation** of **left eye** (lateral rectus muscle is active); inward rotation while trying to depress the eye (unopposed superior oblique muscle); significant neck stiffness.
Labs	N/A
Imaging	CT/MR/Angio: **left posterior communicating artery aneurysm.**
Gross Pathology	N/A
Micro Pathology	N/A
Treatment	Urgent surgery to clip aneurysm.
Discussion	The oculomotor nerve (= CN III) innervates the levator palpebrae superioris; the medial, superior, and inferior recti; and the inferior oblique muscles. CN III palsy may be due to aneurysm, increased intracranial pressure, uncal herniation, diabetes mellitus, hypertension, or giant cell arteritis. Subarachnoid hemorrhage is most commonly due to a **ruptured intracranial aneurysm,** arteriovenous malformation, cerebrovascular accident, or trauma. The **middle cerebral artery bifurcation is the most common location** of intracranial aneurysms.

. . . .

SUBARACHNOID HEMORRHAGE WITH CN III PALSY

ID/CC	A 47-year-old **male** comes to the clinic for evaluation of **loss of sensation** in his lower extremities as well as **lack of sense of position and vibration**; he also has difficulty walking normally.
HPI	He also describes **sharp pain** and numbness (= PARESTHESIAS) in his lower extremities. On directed questioning, he states that about 20 years ago he had a painless ulcer in the glans penis that disappeared spontaneously (= SYPHILITIC PRIMARY CHANCRE).
PE	**Gait wide-based; hyporeflexia and poor muscle tone;** left knee swollen and deformed.
Labs	VDRL positive; FTA-ABS positive.
Imaging	XR: foot cartilage destruction with bone remodeling and overgrowth (= NEUROPATHIC JOINTS OF CHARCOT).
Gross Pathology	Grayish discoloration of posterior nerve roots; thickening of pia mater (meningeal layers covering the spinal cord are the pia, arachnoid, and dura).
Micro Pathology	Degeneration of fasciculus gracilis and posterior root ganglia with proliferation of neuroglia.
Treatment	Penicillin.
Discussion	A late, tertiary manifestation of **syphilis; mainly involves the lumbar spinal cord** (neurosyphilis). On transverse section, the spinal cord has H-shaped gray matter (two posterior horns, two anterior horns, and an intermediate substance); it is surrounded by white matter, which is divided by the anterior median fissure and the posterior median fissure. On each side of the posterior median fissure there is a posterolateral fissure dividing the area between the median sulcus (fissure) and the posterior horn in two—the fasciculus gracilis medially and the fasciculus cuneatus (= BURDACH'S) laterally. **FIRST AID** p.236

TABES DORSALIS

ID/CC	A 55-year-old **woman** complains of intermittent bouts of **excruciating stabbing pain** on the left side of her **face** between the upper lip and lower eyelid (region covered by the maxillary branch of CN V); the pain is so severe that the patient has considered suicide.
HPI	She first experienced this pain two months ago, while she was **chewing** gum. She also reports that **cold drafts** trigger the attacks.
PE	HEENT exam normal; no lymphadenopathy found; no neurologic abnormalities; no tenderness of affected region (vs. sinus infection or other inflammation).
Labs	N/A
Imaging	MR: occasionally shows thickened enhancing trigeminal nerve.
Gross Pathology	N/A
Micro Pathology	N/A
Treatment	Carbamazepine, phenytoin, alcohol injection of nerve, surgical exploration.
Discussion	Also known as "tic douloureux." **If present in a young individual, multiple sclerosis** should be suspected. Although imaging studies usually yield no positive findings, surgical exploration of the posterior fossa in patients who do not respond to medical therapy frequently reveals aberrant blood vessels pressing on nerve root (amenable to decompression).

· ·

TRIGEMINAL NEURALGIA

ID/CC	A 69-year-old male presents with a persistent headache, increasing clumsiness, and frequent bouts of **nausea and vertigo** as well as **difficulty swallowing** (= DYSPHAGIA).
HPI	He has no history of trauma, vaccination, fever, or exanthem. His family reports that he also has **difficulty articulating some words** (= DYSARTHRIA).
PE	VS: normotension. PE: alert and oriented; mild **hoarseness** (due to ipsilateral vocal cord paralysis) with some difficulty swallowing oral secretions (CN IX, X); left side of face reveals **ptosis, miosis, and anhidrosis** (= HORNER'S SYNDROME); pronounced bilateral **nystagmus** in all directions of gaze (CN VIII); decreased sensitivity to light touch and pain on left side of face (CN V); left-sided incorrect appreciation of distance in finger-to-nose movements (= DYSMETRIA); impaired pain sensation in right (contralateral) side of body (spinothalamic tract).
Labs	N/A
Imaging	MR-Brain: infarction in left lateral medullary area. Angio: left **posterior inferior cerebellar artery** occlusion.
Gross Pathology	N/A
Micro Pathology	N/A
Treatment	Treat causative factors (atherosclerosis, emboli) to prevent further damage; rehabilitation.
Discussion	Also known as the syndrome of posterior inferior cerebellar artery occlusion and lateral medullary syndrome, it results from **occlusion of the vertebral artery or its branches** (posterior inferior cerebellar) **to the lateral medulla.** Findings are consistent with involvement of structures that lie in the territory of its distribution: the dorsolateral quadrant of the medulla.

. . . .

WALLENBERG'S SYNDROME

ID/CC	A 57-year-old right-handed male is brought to the emergency room by his relatives because they noticed that although he speaks fluently, he has begun to **use inappropriate words and phrases** to refer to ordinary objects and events in his daily life.
HPI	He suffers from chronic **hypertension** that has been treated with calcium channel blockers.
PE	VS: **hypertension** (BP 170/120). PE: confusion; constructional neurologic deficit; remainder of PE unremarkable.
Labs	EEG: abnormal brain activity in left temporal lobe over supramarginal and marginal gyri.
Imaging	CT/MR: infarct in left posterior cerebral artery territory.
Gross Pathology	N/A
Micro Pathology	N/A
Treatment	Supportive. Speech therapy.
Discussion	Wernicke's ("receptive") aphasia is a disorder of speech that is due to a lesion in the **superior temporal gyrus** that occurs with posterior cerebral artery occlusion. Speech is **fluent but nonsensical,** and there is also an inability to understand spoken or written language. Nonfluent ("expressive") aphasia (= BROCA'S APHASIA) is characterized by the inability to form words; patients know what they want to say but are unable to do so. Whatever they do say, however, is appropriate and meaningful. This disorder results from a lesion in the inferior frontal gyrus.

. . . .

WERNICKE'S APHASIA

ID/CC	A **newborn** female presents with a nonpulsatile **mass** on the left side of her head; an intern is called to address the mother's concerns.
HPI	The intern has been trying to reassure the child's mother that her child's condition is benign, but his reassurances have been to no avail. The delivery was uneventful except for a prolonged expulsive period.
PE	Newborn female in no acute distress; no cyanosis or pallor; no signs of cardiopulmonary involvement; scalp exam discloses skin discoloration with **ecchymosis** and diffuse swelling of soft tissue in left parietal area involving most of the left parietal bone and one-third of the right parietal bone (**crosses suture lines**).
Labs	CBC/Lytes: normal.
Imaging	XR-Skull: no fracture or overlapping of bones at sutures; large scalp mass unrelated to skull clearly lies across sutures.
Gross Pathology	N/A
Micro Pathology	N/A
Treatment	Observation.
Discussion	Caput succedaneum refers to a **benign edema of the soft tissues of the head during delivery**; it characteristically **crosses the midline and cranial sutures** and is not associated with an underlying fracture. The differential diagnosis includes cephalhematoma, which does not present with ecchymosis, is sometimes associated with a fracture, does not cross suture lines (subperiosteal bleeding is limited to one bone), and may take several hours to become evident (caput is immediately seen).

CAPUT SUCCEDANEUM

ID/CC	A 31-year-old white female comes to her family physician for **a routine physical examination.**
HPI	Her medical history is unremarkable. She has been on birth control pills for the past six months.
PE	Three-centimeter, fusiform **fluid-filled submucosal mass along lateral wall of vagina** on speculum exam; no discharge; cervix normal; on palpation, no pain or mobilization; no pelvic masses on bimanual palpation; rectal exam normal.
Labs	Routine lab exams normal; pregnancy test negative.
Imaging	US: translabial approach shows 3-cm cyst in vagina.
Gross Pathology	N/A
Micro Pathology	Simple cyst with serous fluid, lined with a single layer of columnar epithelium.
Treatment	Surgical excision if large or symptomatic.
Discussion	Most vaginal cysts larger than 2 cm are Gartner's duct cysts, which are of mesonephric (= WOLFFIAN) duct origin (due to incomplete closure during embryonic life) and are found along the anterolateral aspect of the vulva or vagina. They may also be found in the broad ligament.

GARTNER'S DUCT CYST

ID/CC	A 22-year-old woman who is in late labor requests **anesthesia** because she has now given up on a "natural birth" delivery (without obstetric anesthesia-analgesia).
HPI	The obstetrician in charge decides to perform a pudendal block.
PE	Head of baby already on perineum, and mother is having contractions every five minutes; fetal heart rate 140/min with no apparent distress; doctor **identifies ischial spine** with index finger and injects a needle through sacrospinous ligament between baby's head and vagina; applies anesthetic (after ensuring that the needle has not pierced a pudendal vessel with risk of hematoma formation) in vicinity of each ischial spine (= TRANSVAGINAL PUDENDAL NERVE BLOCK).
Labs	Prenatal lab studies within normal limits, including coagulation tests.
Imaging	N/A
Gross Pathology	N/A
Micro Pathology	N/A
Treatment	Pudendal nerve block done (complications may include hematoma formation, systemic toxicity when injected intravascularly, and localized infections).
Discussion	The pudendal nerve provides both motor and sensory innervation to the **perineal region;** it passes out of the pelvis through the greater sciatic foramen, wraps around the external surface of the ischial spine, and enters the pelvis again through the lesser sciatic foramen (crossing the sacrospinous ligament). The nerve travels within the fascia of the internal obturator (= PUDENDAL OR ALCOCK'S CANAL) and splits into three terminal branches (perineal nerve, inferior rectal nerve, and dorsal nerve of clitoris).

· ·

PUDENDAL NERVE BLOCK

ID/CC	A 33-year-old female comes to the emergency room with **sudden-onset left lower abdominal pain** together with nausea and vomiting; she **passed out** near the front door of the hospital.
HPI	Her **last menstrual period was 60 days ago** (she has been regular and has never missed a period). She also has a history of **pelvic inflammatory disease**.
PE	VS: **hypotension; tachycardia.** PE: cold, clammy skin and marked **pallor** (hypovolemia); on palpation, abdomen shows muscle spasm and guarding as well as tenderness of left iliac fossa; pelvic exam cannot be performed because of **excessive pain**; needle puncture of posterior cul-de-sac via vagina (= CULDOCENTESIS) shows free intraperitoneal **nonclotting blood** (due to rupture).
Labs	CBC: hematocrit low; mild leukocytosis. **Pregnancy test positive.** UA: normal.
Imaging	US: ectopic pregnancy in ampullary region of the left fallopian tube with **echogenic fluid in cul-de-sac.**
Gross Pathology	Gestational sac with trophoblast in ampullary region of left fallopian tube.
Micro Pathology	N/A
Treatment	Surgical exploration and hemostasis.
Discussion	Ectopic pregnancy refers to extrauterine locations of the fetus; it may be **tubal, abdominal,** or intraligamentous (broad ligament). Risk factors for ectopic pregnancy include **pelvic inflammatory disease, prior ectopic pregnancy, tubal pelvic surgery,** and exposure to teratogens (e.g., DES). In order of frequency, tubal pregnancies commonly occur at the **ampulla, isthmus, fimbriae, or interstitium.** The blood supply to the tubes comes from the ascending branches of the uterine artery, a branch of the internal iliac artery, and the ovarian artery, a direct branch of the aorta (brisk bleeding).

. . . .

RUPTURED ECTOPIC PREGNANCY

ID/CC	A 54-year-old nurse complains of a heavy sensation in her lower abdomen that **worsens when she lifts heavy objects,** together with back pain and increased frequency of urination with a burning sensation (due to altered location of bladder, subsequent stagnation of urine, and thus bacterial proliferation).
HPI	She has given birth to **five children,** all by vaginal delivery. She complains of **urine leakage while coughing, sneezing, or running** (= STRESS INCONTINENCE). Her menses are irregular, but she has otherwise been in good health.
PE	**Downward bulging of anterior wall of vagina** (= CYSTOCELE) with loss of urethrovesical angle exacerbated by straining; protrusion of cervix (= PROLAPSE OF UTERUS).
Labs	N/A
Imaging	Voiding cystourethrogram demonstrates bladder dropping below symphysis pubis during voiding and loss of urethrovesical angle.
Gross Pathology	N/A
Micro Pathology	N/A
Treatment	Bladder resuspension surgery.
Discussion	Usually a result of stretching of pelvic supporting structures during delivery, coupled with years of gravitational weight and menopausal loss of muscle tone. Pelvic floor support is given by the levator ani muscle and its fascia, which continues with the urogenital diaphragm, endopelvic fascia, and cardinal and uterosacral ligaments.

. . . .

UTERINE PROLAPSE WITH CYSTOCELE

ID/CC A 35-year-old ice-hockey player is brought to the emergency room after suffering a violent **blow to his shoulder** during a game; his right **arm hangs** noticeably **lower** than the left and there is a pronounced **bulge** (the clavicle sticks out) at the tip of his shoulder.

HPI Video replay of the game reveals that the patient was bending forward when an opponent speared into the **superior portion of the patient's acromion.**

PE Patient in pain; shoulder has fallen away from clavicle (due to weight of arm); no loss of sensation in arm; on palpation, tenderness in acromioclavicular and coracoclavicular joints; when external end of clavicle is pressed, it returns to original site (= PIANO KEY SIGN).

Labs N/A

Imaging XR-Shoulder (stress view): 10-pound weight suspended from patient's wrist causes marked separation at acromioclavicular joint with acromial depression.

Gross Pathology N/A

Micro Pathology N/A

Treatment Reduction and immobilization; surgery necessary only if coracoacromial ligament ruptured (grade III) or if patient has persistent shoulder pain.

Discussion Also known as a shoulder separation; may be complete or partial. The acromioclavicular ligament prevents anterior-posterior displacement of the clavicle, while the coracoclavicular ligament prevents vertical displacement of the clavicle.

· ·

ACROMIOCLAVICULAR DISLOCATION

ID/CC	A 16-year-old obese male comes to the emergency room from playing basketball because of acute **pain and swelling of his right ankle.**
HPI	While playing, he landed awkwardly on his right foot, which was inverted, producing immediate acute pain, inability to walk, and swelling.
PE	Right ankle joint **swollen with ecchymosis** in most of lateral side of foot; **acutely painful to touch,** mostly underneath fibular end; no bony crepitus felt; distal temperature, pulses, and sensation normal.
Labs	N/A
Imaging	XR-Ankle: no fracture; lateral malleolar soft tissue swelling.
Gross Pathology	N/A
Micro Pathology	N/A
Treatment	Immobilization until pain and swelling subside (usually 3–5 days).
Discussion	Ankle sprains are the most common type of sprain in the body and are usually undertreated (i.e., the length of immobilization time is often too short), with frequent recurrences; with each new sprain, the ligaments become weaker. The ankle joint is held in place and is protected from inversion stresses by the lateral collateral ligament complex, which consists of the anterior talofibular ligament, the calcaneofibular ligament, and the posterior talofibular ligament. On the medial side there is the wide, broad deltoid ligament, which confers protection from eversion stresses. **The anterior talofibular ligament is the most common ligament injured in ankle sprains** and is **secondary to a hyperinversion injury** when the foot is **plantarflexed.**

. . . .

ANKLE SPRAIN

ID/CC	A 35-year-old computer programmer complains of severe, **persistent leg pain** after beginning an intense fitness program involving long-distance running and weight lifting.
HPI	The pain is on the **anterolateral aspect of** the **right leg, radiating from just below the knee to the ankle.**
PE	Poor muscle tone; pain corresponds to anterior compartment of right leg; **region is swollen, tense, and warm; anterior tibial pulse weak** (tearing of muscles, inflammation, and edema with small hemorrhages lead to necrosis); sensory deficit in foot; immediately after exercise, patient has pain on passive extension of toes, with subjective numbness and tingling in foot.
Labs	N/A
Imaging	CT/MR: edema and possible hematoma of anterior compartment muscles.
Gross Pathology	Grayish discoloration of muscle with edema.
Micro Pathology	Ischemia and necrosis of muscle and nerve fibers.
Treatment	Fasciotomy (cutting of fascia) if intracompartmental pressure is > 30 mmHg after exercise.
Discussion	The anterior compartment of the leg has rigid boundaries (tibia, fibula, crural fascia, anterior intermuscular septum) that can trap blood, contribute to increased pressure, and thus lead to an ischemic process. Acute compartment syndrome may be caused by a **crush injury or fracture to the involved extremity.** In this case, the patient has **chronic exertional compartment syndrome,** which can be seen with unusually vigorous exercise. A sequela of forearm compartment syndrome is Volkmann's ischemic contracture, which results in a stiff, nonfunctioning "claw hand" from muscle necrosis and resulting fibrosis.

. . . .

ANTERIOR COMPARTMENT SYNDROME

ID/CC	A 19-year-old gas-station attendant comes to a local clinic complaining of persistent, increasing **pain on the ulnar side of his hand.**
HPI	He was involved in a **fight** with a truck driver two days ago.
PE	Periorbital ecchymosis on left side with no eye involvement; traumatic absence of left upper central incisor; right hand **swollen;** characteristic **depression of head of fifth metacarpal** when looking at fist anteriorly; patient **cannot flex pinkie** because it elicits pain on fifth metacarpal.
Labs	N/A
Imaging	XR: **transverse fracture of fifth metacarpal neck** with palmar angulation.
Gross Pathology	N/A
Micro Pathology	N/A
Treatment	Reduce and apply splint on ulnar side of hand.
Discussion	A common fracture when something hard is **hit with a closed fist.**

ID/CC   A 46-year-old **woman** comes to her family physician complaining of **pain, numbness, and a tingling sensation** (= PARESTHESIAS) on the palmar aspect of her right thumb, her second and third fingers, and the radial side of her fourth finger (fifth finger is always spared); her attacks occur **primarily at night.**

HPI   She has worked for several years in the "data entry" department of a computer firm (an activity associated with prolonged, **repetitive movements of the wrist**).

PE   **Wasting of thenar eminence;** weakness of thumb while opposing to fifth digit (weakness of opponens pollicis); tapping over radial side of palmaris longus tendon produces a tingling sensation (= TINEL'S SIGN); increased sensation (= HYPERESTHESIA) over palmar side of thumb to ring fingers; forced flexion of wrists reproduces symptoms while extension relieves them (= PHALEN'S TEST).

Labs   Nerve conduction studies of median nerve show decreased conduction velocity and increased latency as nerve enters hand.

Imaging   MR: thickening and edema of median nerve or of adjacent tendons.

Gross Pathology   N/A

Micro Pathology   N/A

Treatment   Extension splinting of affected wrist and NSAIDs. Injection of canal with lidocaine and corticosteroids; surgical decompression of transverse carpal ligament (carpal tunnel release) if not responsive to local injections.

Discussion   A type of stenosing tenosynovitis seen in people who use their hands in a repetitive fashion (e.g., those who use computers). The **median nerve** lies between the flexor carpi radialis and flexor digitorum sublimis tendons, and it is covered by the flexor retinaculum. In all, there are nine tendons in the tunnel that compress the nerve against the retinaculum.

. . . .

CARPAL TUNNEL SYNDROME

ID/CC A 55-year-old white male truck driver (whose work involves prolonged abduction of the arms) complains of **tingling, numbness, and pain on the ulnar side** of his left arm and hand.

HPI Three years ago he had a motorcycle accident in which he sustained bruises in the upper thorax, head, and neck.

PE **Diminished left radial pulse** on abduction of arm (cervical rib compresses scalenus anterior muscle and thus subclavian artery) as well as when patient turns head to the left while sitting and holding his breath in inspiration; returns to normal when facing straight (= ADSON'S TEST); **bruit** over left subclavian artery; **percussion over left brachial plexus reproduces symptoms.**

Labs Nerve conduction studies normal.

Imaging XR-Cervical Spine: **short left cervical rib arising from C7.** MR/Angio: narrowing of subclavian artery with **poststenotic dilatation.**

Gross Pathology N/A

Micro Pathology N/A

Treatment Physiotherapy; rib resection in persistent cases.

Discussion Cervical rib syndrome is one of the "**thoracic outlet syndromes,**" as are the scalenus anterior, costoclavicular, and hyperelevation syndromes. These entities compress neurovascular structures and thus give rise to similar signs and symptoms. Cervical ribs of varying sizes are present in a small percentage of the normal population but in some cases may impinge on lower brachial plexus branches or subclavian vessels. **FIRST AID** p.117

· ·

CERVICAL RIB

ID/CC	A 17-year-old high-school student is brought to the emergency room straight from a football game because of acute, severe **pain in the left knee and inability to walk.**
HPI	He could not get back up on his feet after being "chop-blocked" by a lineman **from the side** during a football game.
PE	Leg slightly **flexed;** marked tenderness on medial aspect of knee (damage to medial collateral ligament) and anterior knee joint space (damage to medial meniscus); **positive anterior drawer sign** (ruptured anterior cruciate ligament); **marked knee effusion.**
Labs	Aspiration of affected knee reveals return of grossly bloody fluid (= HEMARTHROSIS).
Imaging	MR-Knee: anterior cruciate tear; torn medial meniscus; rupture of medial collateral ligament; joint effusion.
Gross Pathology	N/A
Micro Pathology	N/A
Treatment	Surgical graft replacement of the anterior cruciate ligament (ACL), suture of medial ligament and repair of medial meniscus.
Discussion	**Unhappy triad (lesion of the ACL, medial meniscus, and medial collateral ligament)** is a common occurrence when the knee receives a blow laterally while the foot is firmly planted and the knee slightly flexed, resulting in massive tension on these structures and tearing. **FIRST AID** p.98

. . . .

COMBINED KNEE INJURY

ID/CC	A 1-week-old child is brought to the pediatrician because his mother noticed that the child **does not move his right arm** normally.
HPI	His delivery was dystopic and prolonged with a **breech presentation.**
PE	Child well nourished and developed; no focal neurologic deficits; right arm **extended, internally rotated, and in adduction; pronation of forearm** (= WAITER'S TIP POSITION).
Labs	Basic workup and newborn screening normal.
Imaging	XR: no fracture of clavicle or arm.
Gross Pathology	N/A
Micro Pathology	N/A
Treatment	Physiotherapy, nerve repair in selected cases.
Discussion	Erb–Duchenne palsy involves the upper brachial plexus (C5–C6), usually by **adduction traction of the arm with hyperextension of the neck.** At times the phrenic nerve may be involved, resulting in ipsilateral diaphragmatic paralysis.

· ·

ERB'S PALSY

ID/CC An **85-year-old woman** is taken to the emergency room after **falling** while climbing out of the bathtub; she has pain in her right groin and **cannot move her right leg.**

HPI She suffers from diabetes, hypertension, and **osteoporosis.** She is currently being treated with calcium supplements and calcitonin.

PE Frail, **elderly woman** with poor muscle tone and low body weight; **right leg externally rotated** at rest (lateral rotators: piriformis, obturator internus and externus, superior and inferior gemellus, quadratus femoris, gluteus maximus); **right leg slightly shorter** than left with tenderness in femoral triangle; limb in **adduction; cannot raise heel off bed.**

Labs N/A

Imaging XR-Plain: intertrochanteric fracture of femur; osteoporosis.

Gross Pathology N/A

Micro Pathology N/A

Treatment Nondisplaced and minimally displaced fractures can be treated by percutaneous pinning vs. screw fixation of the affected hip, while displaced fractures (associated with a high risk of avascular necrosis) are treated with a hip hemiarthroplasty (femoral head prosthesis).

Discussion Frequently seen in elderly postmenopausal women with osteoporosis. The mechanism of fracture is **often a trivial force,** causing subcapital fractures, impacted or not, as well as pretrochanteric, intertrochanteric, or extracapsular fractures. Patients with hip fractures are at high risk for developing deep venous thrombosis postoperatively; thus, proper prophylactic measures (e.g., sequential compression stockings, anticoagulation) must be taken.

. . . .

FEMORAL NECK FRACTURE

ID/CC	A 16-year-old high-school track-team member experiences **intense pain in his right collarbone;** he also notices a **prominence** where the middle third and outer third of his right clavicle meet.
HPI	The patient tripped over a hurdle and **fell on his outstretched hand.**
PE	Marked **tenderness and deformity** at site of fracture; right arm hangs lower than left; arm medially rotated and painful (medial rotators of arm are stronger than lateral rotators).
Labs	N/A
Imaging	XR-Clavicle: fracture of clavicle at middle third; lateral portion depressed.
Gross Pathology	N/A
Micro Pathology	N/A
Treatment	Reunion of fracture by figure-of-eight bandage or clavicular spica cast, with affected arm placed in a sling for comfort, and isolation of arm movement.
Discussion	Fractures are usually located where the middle and outer thirds of the clavicle meet. The medial segment is displaced upward by the sternocleidomastoid muscle, while the distal end is depressed by the weight of the shoulder.

ID/CC	A 44-year-old executive comes to see his physician for **burning pain in his right elbow.**
HPI	He is an avid **tennis player.** For the past several weeks he has experienced a sharp pain in the right elbow (his playing arm) during practice.
PE	When asked to localize pain, patient points to **lateral epicondyle** of humerus; on palpation, area is **warm and exquisitely tender; pain increases with wrist extension against resistance;** no elbow deformity, swelling, or redness; distal pulses normal; no sensory disturbances.
Labs	N/A
Imaging	XR-Elbow: no fracture or dislocation.
Gross Pathology	Union of tendon and underlying periosteum chronically inflamed with tendinitis, synovitis, granulation tissue formation, and bone resorption.
Micro Pathology	Angiofibroblastic proliferation and degenerative fibrosis of the extensor carpi radialis brevis tendon.
Treatment	Discontinue tennis for six weeks; NSAIDs, local heat-cold, topical steroids, surgery in selected cases.
Discussion	The lateral epicondyle of the humerus serves as the origin for the extensors of the wrist: the extensor carpi radialis brevis, extensor digitorum, extensor digiti minimi, and extensor carpi ulnaris. All are innervated by the radial nerve. In this condition, also known as **tennis elbow,** the strain of repeated extension of the wrist against a force, as in playing tennis or throwing a baseball, places considerable stress on the site. Lateral epicondylitis most commonly affects the extensor carpi radialis brevis tendon. Other causes of lateral elbow pain include posterior interosseous nerve compression and radiocapitellar arthritis.

. . . .

LATERAL EPICONDYLITIS

ID/CC   A **6-year-old male** is referred to an orthopedic surgeon by his pediatrician because of the recent onset of a **limp** that has persisted for more than two weeks with no apparent cause.

HPI   He also complains of **pain in the right groin** with radiation to the inner thigh and knee.

PE   Child is well developed and nourished; average weight and height for age; chest and abdomen show no pathology; on palpation over right coxofemoral joint there is **tenderness and muscle spasticity.**

Labs   N/A

Imaging   XR: **small femoral head epiphysis;** sclerosis of flattened femoral head epiphysis. MR: marrow edema and fracture line in femoral head epiphysis. Nuc: abnormal uptake in femoral head.

Gross Pathology   Collapsed, soft, and friable articular cartilage in femoral head.

Micro Pathology   Avascular necrosis of proximal femoral epiphysis.

Treatment   Petrie walking cast (abduction bracing), surgery (femoral versus acetabular osteotomy).

Discussion   A type of **avascular necrosis** that occurs in the femoral heads of children between the ages of 3 and 10 years, affecting males more than females. It is self-limited over a period of up to three years; roughly half of affected children will have residual deformity.

· ·

LEGG–CALVE–PERTHES DISEASE

ID/CC	A 65-year-old woman complains of progressive **difficulty abducting her arm beyond 45 degrees.**
HPI	She underwent a **mastectomy** four months ago for **breast cancer.** Before her surgery, the patient had full range of motion in her arm.
PE	**Winged scapula** noted when patient pushes against wall (serratus anterior paralyzed by nerve damage and therefore unable to fix scapula against chest wall).
Labs	N/A
Imaging	N/A
Gross Pathology	N/A
Micro Pathology	N/A
Treatment	Physiotherapy.
Discussion	In the course of surgical axillary lymph node dissection during mastectomy, the long thoracic nerve may be injured.

. . . .

LONG THORACIC NERVE INJURY

ID/CC	A 23-year-old cross-country motorcycle racer visits an orthopedic surgeon because of **weakness in his right hand.**
HPI	Six weeks ago, he suffered a fall during training that resulted in an **elbow fracture.**
PE	When patient **flexes his wrist, it deviates to the ulnar side** (due to unopposed action of the flexor carpi ulnaris; the median nerve innervates the flexor carpi radialis, which is responsible for flexing the wrist and deviating it radially); when patient is asked to clasp his hands together, right **index finger cannot be flexed** (= OCHSNER'S TEST) (due to inactivity of the flexor digitorum sublimis); patient **cannot flex thumb** (due to inactivity of flexor pollicis longus) **and cannot oppose thumb** (due to inactivity of opponens pollicis brevis, innervated solely by the median nerve); fourth and fifth fingers are flexed with thumb and index finger extended (= BENEDICTION HAND).
Labs	N/A
Imaging	XR: healing of elbow fracture.
Gross Pathology	N/A
Micro Pathology	N/A
Treatment	Physiotherapy; surgical exploration if nerve compression by fracture callus is suspected.
Discussion	The median nerve innervates the flexors at the wrist and fingers as well as the forearm pronators. The ulnar nerve innervates the flexor carpi ulnaris, which, in addition to flexing the wrist, also deviates it medially.

. . . .

MEDIAN NERVE PALSY (NONCARPAL)

ID/CC	A 13-year-old male is referred to an orthopedic surgeon by his pediatrician because of persistent **swelling and pain below the knee** that is intermittent in nature.
HPI	He is a soccer fan and has been **playing soccer every afternoon** in anticipation of the upcoming World Cup. The patient also reports increased pain just below the knee while going up and down stairs.
PE	Athletic-looking, fit teenager; **exquisitely painful** area of **swelling** 4 cm below knee joint (**tibial tuberosity**); when patient extends leg against resistance, pain is elicited or increased.
Labs	N/A
Imaging	XR-Knee: slight avulsion of tibial tubercle with osseous resorption and new bone formation, resulting in a fragmented appearance of the secondary ossification center.
Gross Pathology	N/A
Micro Pathology	N/A
Treatment	Avoid activities that place pressure on area directly or axially, such as kneeling and jumping. Hamstring stretching and ice massage are helpful.
Discussion	The patellar tendon inserts in the anterior tibial tuberosity, which by the end of puberty has an ossification center. With repeated trauma, the tibial tubercle is avulsed and cut off from the blood supply, suffering **avascular necrosis.** The course is **self-limited,** but in some patients a painful bony fragment remains with nonunion.

. . . .

OSGOOD–SCHLATTER'S DISEASE

ID/CC	A 17-year-old boy is rushed to the nearest emergency room after being involved in a **high-speed motorcycle accident.**
HPI	He was treated for **shock** in the ambulance with crystalloids, pressors, and oxygen.
PE	VS: tachycardia; hypotension; no fever. PE: patient in acute distress; respiratory sounds heard in both lung fields; peritoneal lavage negative; **pain** with pressure on **iliac crests** and trochanters bilaterally; **blood present at urethral meatus** (paramedics correctly avoided inserting a Foley catheter); abdomen shows no peritoneal signs.
Labs	CBC: hemoglobin and hematocrit low; leukocytosis. UA: high specific gravity; hematuria.
Imaging	XR-Pelvis: fracture of pelvis bilaterally. Retrograde Urethrography: **membranous urethral rupture** with extravasation of contrast media outside the peritoneal cavity. CT: large hematoma surrounding region of pelvic fractures.
Gross Pathology	N/A
Micro Pathology	N/A
Treatment	Treat shock. Drain urine collection. Suprapubic cystostomy, delay urethral repair. Pelvic suspension, blood replacement. Unstable pelvic fractures may require emergent stabilization with a pelvic external fixator.
Discussion	Generally, compound lesions are to be expected. Bladder and urethral lesions are common in pelvic fractures. Also, fractures of the pelvis may conceal a large volume of blood.

· ·

PELVIC FRACTURE

ID/CC	A 2-year-old girl is brought to the pediatric emergency room crying with **pain** in the left **elbow.**
HPI	While the child was having a temper tantrum, the father **forcefully pulled her by the hand.**
PE	Child is holding right **forearm in pronation and elbow in flexion;** no swelling, ecchymosis, obvious deformity, or bone crepitus.
Labs	N/A
Imaging	XR (AP and Lateral)-Elbow: no fracture or dislocation; child stopped crying right after lateral view was taken.
Gross Pathology	N/A
Micro Pathology	N/A
Treatment	**Gentle supination** of forearm with elbow in 90 degrees of flexion often reduces subluxation during positioning for lateral-view x-rays.
Discussion	Also called **nursemaid's elbow,** subluxation of the radial head occurs when the extended and pronated arm is pulled, tearing the annular ligament.

RADIAL HEAD SUBLUXATION

ID/CC	A 27-year-old female comes to see the orthopedic surgeon because of **inability to extend her right wrist** and fingers.
HPI	She suffered a **middle third humeral fracture** four weeks ago while snow-skiing.
PE	Wrist on right side is "hanging" (= WRIST DROP); **inability to dorsiflex hand** over forearm; inability to elevate thumb; **anesthesia on dorsum of hand over thumb and first three digits.**
Labs	N/A
Imaging	XR: middle third of right humerus shows transverse fracture with callus formation.
Gross Pathology	N/A
Micro Pathology	N/A
Treatment	Closed reduction and hanging cast or "U"-shaped splint. Wrist immobilization. Surgical radial nerve exploration if nerve conduction studies are abnormal for several months.
Discussion	The radial nerve courses through the spiraling groove located in the middle third of the humeral shaft, predisposing it to lesions. Radial nerve palsy may be delayed by months or even years (due to trapping of the nerve in osseous callus or scar tissue).

. . . .

RADIAL NERVE PALSY

ID/CC A 19-year-old college undergrad comes to an urgent care center at a ski resort because of **pain in the wrist.**

HPI He has suffered repeated falls while snowboarding (falling forward with outstretched palms).

PE **Hyperesthesia** and marked **tenderness in anatomical snuff box** (bounded by the **extensor pollicis longus** and the **extensor pollicis brevis;** the **scaphoid** and **trapezium** bones lie in the floor).

Labs N/A

Imaging XR-Wrist: no definite fracture (a small fracture of the scaphoid may not appear on x-ray for several weeks until the damaged bone in the region is undergoing resorption). Special radiographic views of the scaphoid as well as CT, MR, or nuclear medicine scans may be obtained for diagnosis if strong clinical suspicion exists.

Gross Pathology N/A

Micro Pathology N/A

Treatment Immobilization in a long-area thumb spica cast, which immobilizes the first thumb phalanx until there is radiologic evidence of fracture healing (may take > 10 weeks).

Discussion The scaphoid and lunate bones articulate with the radius. The scaphoid is a boat-shaped carpal bone that has a tubercle; fractures may involve the tubercle, the proximal pole, or the middle third. The fracture often goes unrecognized, and there is a chance of **avascular necrosis,** mainly in displaced proximal pole fractures, since the scaphoid bone, like the talus and femoral head, has a very tenuous blood supply.

· ·

SCAPHOID FRACTURE

ID/CC	During anti-Gulf War protests in Ohio, a 23-year-old man was forcefully **dragged away by the arm** because he was blocking the entrance to the meeting.
HPI	While in the police car on the way to headquarters, he complained of pain in the shoulder and **inability to move his arm.**
PE	Pain in shoulder, **deformity** (lack of normal rounded contour of shoulder), and inability to move arm; **depression** easily palpable under acromion; humeral head palpable through axilla. After reduction of the dislocation, patient complains of **numbness on the lateral aspect of the forearm** and **weakness in biceps muscle function** when compared to uninvolved side.
Labs	Basic lab work normal; no trace of alcohol in blood; no drugs in urine.
Imaging	XR: depression fracture of the posterolateral articular surface of humeral head (= HILL–SACHS LESION); axillary view of glenohumeral joint shows humeral head to be anterior to glenoid fossa.
Gross Pathology	N/A
Micro Pathology	N/A
Treatment	Before reducing the dislocation, one must look for possible neurologic-vascular damage.
Discussion	The glenohumeral joint is frequently dislocated due to its poor osseous stability. Anterior dislocations (the humeral head normally lies in front of the coracoid process of the scapula) usually result from a fall on the arm in forced abduction and extension. Musculocutaneous nerve injury is possible (it supplies the coracobrachialis as well as the brachialis and biceps muscles and provides sensation to the lateral area of the forearm). Posterior shoulder dislocations are much less common and are seen following electric shock injuries and grand mal seizures.

. . . .

SHOULDER DISLOCATION

ID/CC   An 18-year-old female high-school student is brought to the emergency room after **slashing** the palmar side of her **left wrist** at the skin lines of flexion with a razor blade; she is **unable to flex her wrist or oppose her thumb.**

HPI   She had been severely depressed for several months because her boyfriend had been seeing other women.

PE   Moderate bleeding (superficial branch of radial artery); inability to flex wrist (severed tendon of palmaris longus); failure to oppose thumb and anesthesia over thumb and first/second digits (paralysis of thenar muscles and loss of sensation due to **severed median nerve**); thumb abduction still possible by abductor pollicis longus, innervated by radial nerve.

Labs   N/A

Imaging   N/A

Gross Pathology   N/A

Micro Pathology   N/A

Treatment   Surgical hemostasis and repair of severed structures. Psychiatric treatment.

Discussion   The median nerve passes deep to the flexor retinaculum; it innervates the thenar and lumbrical muscles and supplies sensory branches to the lateral palmar surface, the sides of digits 1, 2, and 3, and the lateral side of digit 4. The tendon of the palmaris longus lies medial, parallel, and superficial to the median nerve. The 10 structures that lie within the carpal tunnel include the four flexor digitorum superficialis tendons, the four flexor digitorum profundus tendons, the flexor pollicis longus tendon, and the median nerve. The palmaris longus tendon, ulnar nerve, and ulnar artery all lie volar to the transverse carpal ligament, which forms the roof of the carpal tunnel.

· ·

SLASHING OF WRISTS

ID/CC	A 20-year-old college student complains of **acute, severe pain on both sides of his face** just in front of his ears that began **while yawning** and radiates to both ears; he **cannot close his mouth,** and he is **unable to speak clearly.**
HPI	He was well until the present complaint, and this is the first time these symptoms have occurred.
PE	Protrusion of lower jaw; **dimpling of skin in temporomandibular joint** area.
Labs	N/A
Imaging	XR: mandibular condyles dislocated anteriorly from temporal fossa.
Gross Pathology	N/A
Micro Pathology	N/A
Treatment	Grip mandible firmly in hands with thumbs placed behind second molar; push mandible inferiorly and posteriorly in quick, single motion (= MANUAL REDUCTION).
Discussion	Clinical diagnosis is usually made on the spot. The physician may elect to correct dislocation under general anesthesia, but this is generally unnecessary.

ID/CC A mother brings her 3-year-old child to an orthopedic surgeon for evaluation of his gait; she says **the child "waddles"** when he walks.

HPI The child had difficulty learning how to walk and has always been rather unstable in his gait.

PE When patient stands on his left leg, his right buttock sags (= TRENDELENBURG SIGN); no sensory loss noted in gluteal area; swing phase of left leg seems most affected; to swing left leg, child **leans over to right side** and then swings left leg in front of right (the **superior gluteal nerve** is paralyzed on the right side of this child); right leg swings normally (hip abductors function normally on the left side to prevent pelvis from tilting over to the right side when right leg is swinging).

Labs Screen for inherited metabolic diseases negative; basic lab work normal.

Imaging XR: right hip dislocation.

Gross Pathology N/A

Micro Pathology N/A

Treatment Walking stick or cane in left hand to prevent hip from tilting over to left side when left leg is swinging. Surgery.

Discussion Unilateral hip dislocation causes Trendelenburg's gait, with **tilting of the trunk toward the affected side** in each step. The superior gluteal nerve exits the sciatic foramen superior to the piriformis muscle. This nerve innervates the gluteus medius, gluteus minimus, and tensor fascia lata, which are medial rotators of the thigh and abductors of the hip when the thigh is fixed. The inferior gluteal nerve and artery as well as the sciatic nerve exit the sciatic foramen inferior to the piriformis muscle to supply the gluteus maximus.

. . . .

TRENDELENBURG GAIT

ID/CC	A 73-year-old male is referred to a surgeon because of a painful **mass in the left inguinal area**; the mass protrudes with straining and **disappears at rest**.
HPI	The patient has frequent bouts of **constipation** (resulting in increased intra-abdominal pressure).
PE	Rectal exam shows marked prostatic enlargement (straining at micturition is predisposing factor); in supine position, **left inguinal region** does not disclose any apparent pathology, but when patient stands and is asked to cough, **a mass** is felt in the inguinal canal that can be easily reduced.
Labs	N/A
Imaging	N/A
Gross Pathology	N/A
Micro Pathology	N/A
Treatment	Surgical repair; treatment for prostatic hyperplasia comes first.
Discussion	Direct hernias are more frequently seen in elderly people with weak abdominal wall musculature; they protrude **directly through the floor of the muscular inguinal canal**, which is the transversalis fascia (inside Hesselbach's triangle). On scrotal examination, with the finger introduced in the inguinal canal through the external inguinal ring, the hernia is felt in the pulp of the finger (indirect hernias are felt in the tip of the finger). The limits of Hesselbach's triangle are the deep (inferior) epigastric artery, lateral rectus muscle border, and inguinal ligament. Direct hernias may contain urinary bladder that can be damaged during surgery. **FIRST AID** p.97

ID/CC	A **1-day-old** male infant (vs. pyloric stenosis, which is associated with 3-week-old infants) has persistent **bilious vomiting** (vs. pyloric stenosis, which is nonbilious) when his mother attempts to feed him.
HPI	He was born at 36 weeks of gestation and weighed 2.3 kg at birth (vs. full-term for pyloric stenosis). His mother had **excess amniotic fluid** (= POLYHYDRAMNIOS).
PE	VS: tachycardia. PE: dehydrated; no jaundice present; on chest auscultation, continuous machinery murmur (patent ductus arteriosus) present; single palmar crease and typical facies of Down's syndrome; **abdomen distended,** tympanic to percussion, and painful; no mass felt; no visible peristalsis; normal-looking meconium on rectal exam.
Labs	N/A
Imaging	KUB: gaseous dilatation of stomach and duodenum (= "DOUBLE BUBBLE").
Gross Pathology	N/A
Micro Pathology	N/A
Treatment	Surgical repair.
Discussion	Duodenal atresia is associated with other musculoskeletal and visceral abnormalities, notably **Down's syndrome,** as well as with congenital heart disease. It typically presents with **bilious vomiting** on the **first day of life.**

DUODENAL ATRESIA

ID/CC	A 28-year-old male weight lifter comes to the emergency room because of a **painful lump in the right scrotal area** that began earlier in the morning.
HPI	He is healthy with an unremarkable past medical history; he frequently engages in strenuous physical labor.
PE	VS: tachycardia. PE: abdomen distended and tympanic with increased peristaltic movements; tender to palpation with no rebound tenderness (intestine is obstructed with backward accumulation of gas and feces); **tender, tense, and painful mass** in right inguinal area **that continues** through external inguinal ring **into scrotum;** mass does not transilluminate.
Labs	CBC: neutrophilic leukocytosis (intestinal loop is suffering ischemia). Increased BUN; normal creatinine (dehydration due to intra-abdominal sequestration).
Imaging	KUB: dilated small bowel loops with air–fluid levels in stepladder pattern; mass in right scrotum.
Gross Pathology	N/A
Micro Pathology	N/A
Treatment	Surgical treatment to free intestinal loop and repair hernia.
Discussion	Indirect inguinal hernia is the **most common type of hernia** in males (usually young) and females. The hernia comes out through the internal inguinal ring with the spermatic cord and frequently protrudes into the scrotum through the external inguinal ring. Indirect hernias lie **lateral-superior to the inferior epigastric artery, outside Hesselbach's triangle.** Indirect inguinal hernias are most commonly due to a **congenitally patent processus vaginalis.** **FIRST AID** p.97

INDIRECT INGUINAL HERNIA

ID/CC	A 75-year-old man complains of **difficulty swallowing** (= DYSPHAGIA) and **speaking** (= DYSARTHRIA) (due to compression of CN IX, X); these symptoms have progressively worsened over the past several months.
HPI	He has also noticed increasing pain during urination (= DYSURIA) over the past several months and **difficulty starting the flow of urine** (a consequence of prostatic carcinoma).
PE	Weakness of right pharyngeal and laryngeal muscles; **atrophy of sternocleidomastoid muscle and trapezius muscle** (compression of CN XI); rectal exam reveals rock-hard, fixed mass in prostate.
Labs	Markedly increased serum-specific prostatic antigen; increased acid phosphatase and alkaline phosphatase.
Imaging	CT-Neck: **mass in jugular foramen.**
Gross Pathology	Metastatic prostate carcinoma impinging on **CN IX, X, XI at level of jugular foramen.**
Micro Pathology	Transrectal biopsy shows high-grade prostatic adenocarcinoma.
Treatment	Prostate carcinoma treated by orchiectomy and hormonal modalities; radiation.
Discussion	Prostatic carcinoma produces metastases in the axial skeleton with the possibility of involvement at all levels of the spine as well as the cranial bones.

. . . .

MASS IN JUGULAR FORAMEN

ID/CC	A **2-year-old** boy is brought to the emergency room by his parents because of an **increase in the size of his belly** and persistent **vomiting.**
HPI	Two weeks ago the boy had **bright red blood in his stools** for four days.
PE	VS: tachycardia. PE: **pallor** in conjunctiva; abdomen **distended and tympanic** with increased bowel sounds; on palpation, abdomen is tender with small, **sausage-shaped mass** in right lower quadrant (due to intussusception).
Labs	CBC: normochromic, normocytic anemia; neutrophilic leukocytosis. Increased BUN; creatinine normal.
Imaging	KUB: air–fluid levels with small bowel loop distention. Nuc: presence of **ectopic gastric mucosa** confirmed.
Gross Pathology	Five-centimeter-long diverticulum situated on antimesenteric border of ileum located 60 cm from ileocecal valve. Diverticulum forms tip of an intussusception.
Micro Pathology	Contains **ectopic acid-secreting gastric mucosa** and pancreatic tissue.
Treatment	Surgical excision.
Discussion	The most common congenital anomaly of the GI tract; consists of a diverticular sac caused by **persistence of the vitelline duct** or yolk stalk. The five 2's describe it: 2 inches long, 2 feet from the ileocecal valve, 2% of the population, first 2 years of life, 2 types of epithelium. May be asymptomatic or may give rise to intussusception and intestinal obstruction, diverticulitis (indistinguishable from appendicitis), or bleeding. **FIRST AID** p.95

MECKEL'S DIVERTICULUM

ID/CC	A 2-week-old **male** is brought to the family doctor because his parents noticed a "**lump near the child's buttocks**"; the lump sometimes disappears but invariably reappears when the child **cries**.
HPI	He is the first-born child of a healthy Hispanic couple. The pregnancy and delivery were uneventful.
PE	VS: no fever. PE: no abdominal masses palpable; no neurologic signs; upon examination of **left lumbar area,** nothing was noticed until child cried, at which point a 2-cm-diameter, **rounded mass** was felt on edge of iliac crest.
Labs	Basic lab work normal; screening for inherited metabolic deficiencies normal.
Imaging	N/A
Gross Pathology	N/A
Micro Pathology	N/A
Treatment	Surgical.
Discussion	Petit's triangle is formed by the iliac crest inferiorly, the posterior border of the external oblique anteriorly, and the anterior border of the latissimus dorsi muscle posteriorly. Petit's triangle hernias are seen in all age groups and are more common in males, arising more **frequently on the left side.**

PETIT'S TRIANGLE HERNIA

ID/CC	A 33-year-old gas station attendant is brought to the emergency room after sustaining a **bullet wound on the back of his leg.**
HPI	He panicked and ran when confronted by two muggers on a dark, deserted street (shot from behind).
PE	Left foot is **cold; inability to dorsiflex foot;** overlying skin **cyanotic; no dorsalis pedis pulse** palpable (artery lies between tendons of extensor hallucis longus and extensor digitorum longus); entry wound located on popliteal fossa and exit wound on anterolateral portion of knee; lower third of thigh and upper third of leg tense, swollen, and painful.
Labs	N/A
Imaging	XR: comminuted (multiple fragments) fracture of tibial plateau. Angio: **traumatic transection of popliteal artery.**
Gross Pathology	N/A
Micro Pathology	N/A
Treatment	Emergency surgical repair of artery.
Discussion	The popliteal fossa is a diamond-shaped zone bounded on the lateral superior margin by the biceps femoris muscle, the medial superior margin by the semimembranous muscle, and the inferior margins by the gastrocnemius muscle. It harbors the popliteal artery and vein (where the lesser saphenous vein drains), tibial nerve, peroneal nerve, and obturator and femorocutaneous nerve branches. **Traumatic knee dislocations** are associated with **intimal tears of the popliteal artery.**

ID/CC   A 37-year-old female comes to the emergency room because of the sudden development of **pain in the right groin** area and a **tender mass** of two hours' duration.

HPI   She is a healthy mother of four with no pertinent medical history.

PE   VS: tachycardia; mild hypertension (due to pain). PE: patient in pain; chest and abdomen normal; **no signs of intestinal obstruction; small,** rounded, very tense and tender **mass** felt; on palpation of right groin area, mass is not reducible and causes **intense pain.**

Labs   CBC: marked neutrophilic leukocytosis.

Imaging   KUB: normal appearance of gas in rectum; no signs of intestinal obstruction.

Gross Pathology   N/A

Micro Pathology   N/A

Treatment   Immediate surgery to release ischemic-gangrenous bowel and hernia repair.

Discussion   Richter's hernia refers to a type of hernia in which **only one wall of the intestine** (usually the antimesenteric border) is trapped by the constriction ring of the hernia; it can occur with femoral, inguinal, or umbilical hernias (more common in the **femoral type** because of the narrow orifice). Since gas and feces may still pass through the nonconstricted area, signs of obstruction are usually absent. Femoral (crural) hernias, found medial to the femoral vein in the femoral canal, are more common on the right side, more common in women, and prone to strangulate early. The femoral ring is formed by the inguinal ligament, the lacunar (= GIMBERNAT'S) ligament, the femoral vein (easily damaged during repair), and the pelvic border.

. . . .

RICHTER'S HERNIA

ID/CC	A 23-year-old male is brought to the emergency room in a confused state after being involved in a high-speed downhill **skiing accident.**
HPI	He complains of severe **abdominal pain radiating to the left scapula.**
PE	VS: **marked hypotension** (BP 70/50). PE: cold, clammy skin; acute distress; generalized abdominal tenderness and rebound tenderness with guarding, especially in left upper quadrant; pain in left scapula when foot of bed is elevated or on palpation of left subcostal region (= KEHR'S SIGN) (due to presence of free intraperitoneal blood that irritates diaphragm); dullness to percussion on left flank and dullness to percussion on right flank that changes with position (= BALLANCE'S SIGN).
Labs	Low hematocrit; **peritoneal tap grossly bloody.** UA: negative.
Imaging	CT/US: hematoma surrounding spleen with obliteration of splenic outline; peritoneal fluid.
Gross Pathology	N/A
Micro Pathology	N/A
Treatment	Immediate blood and volume replacement; emergency surgical spleen removal (= SPLENECTOMY) or, when possible, splenorrhaphy (= SPLENIC SUTURE). Postsplenectomy patients should receive pneumococcal vaccine for prophylaxis.
Discussion	The spleen is the most commonly injured organ in blunt abdominal trauma. Sometimes a few days will pass between a trauma and symptomatology (= DELAYED SPLENIC RUPTURE). With splenic enlargement, even minor trauma can cause rupture.

. . . .

RUPTURE OF SPLEEN

ID/CC	A 70-year-old **hypertensive male** was brought to the emergency room because of the **sudden** development of **severe, tearing abdominal pain that radiated to the back.**
HPI	He **lost consciousness** when he was being transported to the hospital in his neighbor's car.
PE	VS: **hypotension** (BP 70/30); tachycardia (HR 110); marked tachypnea. PE: confused, disoriented, and in a **delirious** state; skin cold and clammy; peribuccal cyanosis; **pulsatile mass in abdomen;** while central lines were being placed, patient suffered a fatal cardiac arrest.
Labs	N/A
Imaging	CT-Abdomen: diagnostic; arteriography useful for planning surgical treatment.
Gross Pathology	Autopsy showed a 10-cm-diameter aneurysmal dilatation of abdominal aorta (normal diameter is 2 cm) with abundant atherosclerosis of the wall and rupture.
Micro Pathology	Atherosclerosis (Marfan's syndrome patients show cystic necrosis of the tunica media).
Treatment	Immediate surgical resection and grafting. Patients with asymptomatic abdominal aortic aneurysms > 5 cm usually undergo elective surgical resection.
Discussion	**Atherosclerosis** is the most common cause of abdominal aortic aneurysm (localized dilatation of its lumen). Other causes include syphilis and trauma. Aneurysms are most common in males, particularly the elderly. They are usually located below the level of the renal arteries.

· ·

RUPTURED ABDOMINAL ANEURYSM

ID/CC	A 73-year-old female comes to the emergency room complaining of **acute abdominal pain** that is colicky in nature, along with pain in between contractions and **inability to pass flatus.**
HPI	She suffers from **chronic constipation** and takes laxatives every day.
PE	VS: **tachycardia** (HR 97); borderline hypertension (BP 140/95); fever (38.1 C). PE: **dehydration** with dry mucous membranes; abdomen **markedly distended** and painful with generalized tympanic tone on percussion and absence of peristaltic movements.
Labs	CBC: neutrophilic **leukocytosis.** Increased BUN; normal creatinine; amylase mildly elevated. UA: increased specific gravity.
Imaging	KUB: **massive distention of sigmoid colon.** BE: **bird's-beak** appearance of contrast at point of volvulus.
Gross Pathology	Sigmoid excessively mobile and twisted over its own mesentery with massive distention and thinning (paperlike quality) of intestinal wall.
Micro Pathology	N/A
Treatment	Sigmoidoscopic decompression; with failure, surgical operation.
Discussion	Volvulus (twisting) of the colon is more frequent in the elderly and most commonly occurs in the sigmoid area; the second most common site is the cecum. Closed-loop intestinal obstruction ensues and, if persistent, may lead to gangrene and perforation with peritonitis. Bloody colonic discharge with shedding of dark colonic mucosa suggests colonic necrosis and warrants emergent surgical resection of necrotic bowel.

· ·

SIGMOID VOLVULUS

ID/CC A 13-year-old boy is brought to the emergency room by his parents after an accident at school; he was walking along a steel guard rail when **he slipped and fell, straddling the rail.**

HPI He was in extreme pain initially. At home the pain subsided somewhat. Upon urination, a few **drops of bloody urine** were produced. The child also noticed swelling of his scrotum.

PE VS: tachycardia (HR 120); fever (38.3 C). PE: patient in pain; genital examination reveals crusts of **blood on meatus, ecchymosis,** and **painful swelling of scrotum and perineal region;** abdominal exam discloses a rounded, tender enlargement on the suprapubic area that is nonmotile and dull to percussion (bladder is full).

Labs CBC: **leukocytosis** (17,000) with neutrophilia. Slightly increased BUN with normal creatinine. UA was not possible in the ER (due to inability to void; inserting Foley catheter was contraindicated because of probable urethral damage).

Imaging Urethrography: extravasation of urine into scrotal tissue and perineum (rupture of urethra).

Gross Pathology N/A

Micro Pathology N/A

Treatment Surgical repair, temporary cystostomy.

Discussion The male urethra is divided into several portions: the prostatic (widest), membranous (narrowest), bulbous (in the bulb part of the corpus spongiosum), and spongy portions (longest, within the corpus spongiosum itself), ending in the fossa navicularis and meatus. With forceful constriction of the urethra against the pubic arch, a urethral rupture may ensue. The fascial planes in the region direct urine to flow anteriorly to the loose areolar tissue of the scrotum and superficial perineal space. This space lies between the inferior fascia of the urogenital diaphragm and the superficial perineal fascia.

. . . .

STRADDLE INJURY

ID/CC	A 42-year-old female on the gynecology ward complains of a **dull, aching pain on her left flank** as well as nausea and vomiting on her third postoperative day; the intern notices that the patient has also been **oliguric** overnight.
HPI	She recently underwent a total abdominal **hysterectomy** due to large uterine fibroids.
PE	VS: **low-grade fever;** normotension. PE: patient well hydrated; no pallor noted; surgical wound is midline infraumbilical; **dressing noted to be wet with urine;** abdomen has muscle guarding; peristalsis diminished without obvious peritoneal signs.
Labs	CBC: **leukocytosis** with neutrophilia. BUN and creatinine increased. Lytes: normal. UA: proteinuria and microscopic hematuria.
Imaging	Excretory Urography: **blockage of urine** at level of left **ureter** and **intra-abdominal leakage of urine from right ureter.**
Gross Pathology	N/A
Micro Pathology	N/A
Treatment	Exploration and surgical repair of ureters with end-to-end anastomosis or T stent.
Discussion	The ureters may be injured during hysterectomy. The critical point comes during ligation of the uterine vessels at the vicinity of the cervix. The **ureter crosses underneath the uterine arteries** and is thus vulnerable to injury.

SURGICAL URETERAL INJURY

ID/CC A 48-year-old male computer programmer is brought to the emergency room with **intermittent, excruciating pain in the right flank**; the pain **radiates to his inner thigh and testicle** and is accompanied by nausea and vomiting.

HPI Over a period of a few hours, the **pain migrated toward his groin.** It lasted for 30 minutes and then stopped for another 30 minutes before suddenly recurring (due to periodic peristaltic motion of ureter).

PE Abdominal exam shows **no rebound tenderness** (vs. peritonitis or appendicitis); guarding is present; painful and difficult urination (= DYSURIA); blood in urine (= HEMATURIA); patient is restless and keeps switching position (vs. peritonitis, in which patient lies still because of pain).

Labs UA: **hematuria**; bacteriuria; leukocyturia.

Imaging IVP or CT Urogram: filling defects in ureter and renal calyces due to stones. Obstruction proximal to stone.

Gross Pathology N/A

Micro Pathology N/A

Treatment Hydration, analgesics, antispasmodics; many patients pass stones spontaneously; treat metabolic abnormalities; percutaneous stone extraction (= PERCUTANEOUS NEPHROLITHOTOMY), shock wave stone fragmentation (= EXTRACORPOREAL LITHOTRIPSY), or surgical removal.

Discussion Renal tract stones may produce one of the **severest forms of pain** known due to obstruction and smooth muscle contraction. Calculi may be formed of calcium, oxalate, magnesium ammonium phosphate, cystine, or uric acid. Approximately 85% of renal calculi are **radiopaque calcium oxalate stones.** Uric acid stones are radiolucent.

· ·

UROLITHIASIS

ID/CC	A 43-year-old female complains of numbness and swelling of the legs; she also has **muscle fatigue** in the afternoons with a feeling of **heaviness** in the lower extremities associated with cutaneous **lumps and bumps.**
HPI	She is **overweight,** works behind the counter at a fast-food restaurant (associated with prolonged standing), and has **given birth to five** children. Her **symptoms are alleviated when she elevates her legs** by placing them on a chair.
PE	Obese; difficulty breathing after climbing a flight of stairs; facial plethora; examination of lower extremities reveals swelling with **dilatation of veins** in territory of greater saphenous vein.
Labs	CBC: increased hematocrit. Hypercholesterolemia.
Imaging	N/A
Gross Pathology	N/A
Micro Pathology	N/A
Treatment	Lose weight, **elastic stockings**, diminish prolonged standing, surgery (saphenectomy).
Discussion	The greater saphenous vein starts at the anterior aspect of the medial malleolus, originating in the dorsal venous arch of the foot; it ascends in the anteromedial portion of the leg and thigh to drain in the femoral vein just before the inguinal ligament. It has numerous communicating veins with the deep venous system. Primary varicosities are a result of **incompetence of the valves** in the saphenous vein or in the communicating veins, thus increasing hydrostatic pressure and producing dilatation and tortuous veins. Secondary varicosities are a result of obstruction of the deep venous system with resultant increased flow and pressure.

. . . .

VARICOSE VEINS

ID/CC	A 45-year-old man, the father of seven children, comes to a family planning clinic for advice regarding **birth control.**
HPI	After carefully weighing all possible alternatives with the doctor, he decides to have a **vasectomy.**
PE	Physical exam unremarkable except for an old McBurney appendectomy scar; no contraindications for surgery.
Labs	N/A
Imaging	CXR: within normal limits for age.
Gross Pathology	N/A
Micro Pathology	N/A
Treatment	Vasectomy (complications include scrotal hematoma, infection, spermatic granuloma, spermatocele, and spontaneous recanalization).
Discussion	Vasectomy is an increasingly popular method of permanent birth control (regarded as such, although reports of up to 70% successful reversal exist, mostly in men under 30 years of age who underwent the procedure less than seven years ago). The layers to cut through are skin, superficial scrotal fascia (= DARTOS FASCIA), external spermatic fascia, cremasteric fascia and muscle, internal spermatic fascia, preperitoneal fat, and tunica vaginalis. The ductus deferens is tied in two places and transected.

· ·

VASECTOMY

From the authors of *Underground Clinical Vignettes*

A true classic used by over 200,000 students around the world. The '99 edition features details on the new computerized test, new color plates and thoroughly updated high-yield facts and book reviews. Bi-directional links with the *Underground Clinical Vignettes Step 1* series. ISBN 0-8385-2612-8.

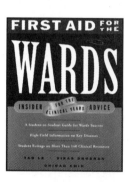

This high-yield student-to-student guide is designed to help students make the transition from the basic sciences to the hospital wards and succeed on their clinical rotations. The book features an orientation to the hospital environment, tips on being an effective and efficient junior medical student, student-proven advice tailored to each core rotation, a database of high-yield clinical facts, and recommendations for clinical pocket books, texts, and references. ISBN 0-8385-2595-4.

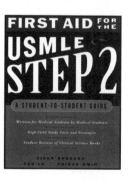

This entirely rewritten second edition now follows in the footsteps of *First Aid for the USMLE Step 1*. Features an exam preparation guide geared to the new computerized test, basic science and clinical high-yield facts, color plates and ratings of USMLE Step 2 books and software. Bi-directional links with the *Underground Clinical Vignettes Step 2* series.

This top rated (5 stars, *Doody Review*) student-to-student guide helps medical students effectively and efficiently navigate the residency application process, helping them make the most of their limited time, money, and energy. The book draws on the advice and experiences of successful student applicants as well as residency directors. Also featured are application and interview tips tailored to each specialty, successful personal statements and CVs with analyses, current trends, and common interview questions with suggested strategies for responding. ISBN 0-8385-2596-2.

The *First Aid* series by Appleton & Lange...the review book leader.
Available through your local health sciences bookstore !

About the Authors

. . . .

VIKAS BHUSHAN, MD

Vikas is a diagnostic radiologist in Los Angeles and the series editor for *Underground Clinical Vignettes*. His interests include traveling, reading, writing, and world music. He is single and can be reached at vbhushan@aol.com

CHIRAG AMIN, MD

Chirag is an orthopedics resident at Orlando Regional Medical Center. He plans on pursuing a spine fellowship. He can be reached at chiragamin@aol.com

TAO LE, MD

Tao is completing a medicine residency at Yale-New Haven Hospital and is applying for a fellowship in allergy and immunology. He is married to Thao, who is a pediatrics resident. He can be reached at taotle@aol.com

HOANG NGUYEN

Hoang (Henry) is a third-year medical student at Northwestern University. Henry is single and lives in Chicago, where he spends his free time writing, reading, and enjoying music. He can be reached at hbnguyen@nwu.edu

JOSE M. FIERRO, MD

Jose (Pepe) is beginning a med/peds residency at Brookdale University Hospital in New York. He was a general surgeon in Mexico and worked extensively in Central Africa. His interests include world citizenship and ethnic music. He is married and can be reached at jmfierro@aol.com

PARAG MATHUR

Parag is a second-year medical student at Mayo School of Medicine. He lives in Rochester, Minnesota, where he remembers his sunny home in Southern California. He can be reached at mathur.parag@mayo.edu